S0-BWW-851

Lecture Notes in Biomathematics

Managing Editor: S. Levin

18

Mathematics and the Life Sciences

Springer-Verlag
Berlin Heidelberg New York

Lecture Notes in Biomathematics

Vol. 1: P. Waltman, Deterministic Threshold Models in the Theory of Epidemics. V, 101 pages. 1974.

Vol. 2: Mathematical Problems in Biology, Victoria Conference 1973. Edited by P. van den Driessche. VI, 280 pages. 1974.

Vol. 3: D. Ludwig, Stochastic Population Theories. VI, 108 pages. 1974.

Vol. 4: Physics and Mathematics of the Nervous System. Edited by M. Conrad, W. Güttinger, and M. Dal Cin. XI, 584 pages. 1974.

Vol. 5: Mathematical Analysis of Decision Problems in Ecology. Proceedings 1973. Edited by A. Charnes and W. R. Lynn. VIII, 421 pages. 1975.

Vol. 6: H. T. Banks, Modeling and Control in the Biomedical Sciences. V, 114 pages. 1975.

Vol. 7: M. C. Mackey, Ion Transport through Biological Membranes, An Integrated Theoretical Approach. IX, 240 pages. 1975.

Vol. 8: C. DeLisi, Antigen Antibody Interactions. IV, 142 pages. 1976.

Vol. 9: N. Dubin, A Stochastic Model for Immunological Feedback in Carcinogenesis: Analysis and Approximations. XIII, 163 pages. 1976.

Vol. 10: J. J. Tyson, The Belousov-Zhabotinskii Reaktion. IX, 128 pages. 1976.

Vol. 11: Mathematical Models in Medicine. Workshop 1976. Edited by J. Berger, W. Bühler, R. Repges, and P. Tautu. XII, 281 pages. 1976.

Vol. 12: A. V. Holden, Models of the Stochastic Activity of Neurones. VII, 368 pages. 1976.

Vol. 13: Mathematical Models in Biological Discovery. Edited by D. L. Solomon and C. Walter. VI, 240 pages. 1977.

Vol. 14: L. M. Ricciardi, Diffusion Processes and Related Topics in Biology. VI, 200 pages. 1977.

Vol. 15: Th. Nagylaki, Selection in One- and Two-Locus Systems. VIII, 208 pages. 1977.

Vol. 16: G. Sampath, S. K. Srinivasan, Stochastic Models for Spike Trains of Single Neurons. VIII, 188 pages. 1977.

Vol. 17: T. Maruyama, Stochastic Problems in Population Genetics. VIII, 245 pages. 1977.

Vol. 18: Mathematics and Life Sciences, Proceedings 1975. Edited by D. E. Matthews. VIII, 385 pages. 1977.

Lecture Notes in Biomathematics

Managing Editor: S. Levin

18

Mathematics and the Life Sciences

Selected Lectures,
Canadian Mathematical Congress,
August 1975

Edited by David E. Matthews

Springer-Verlag
Berlin Heidelberg New York 1977

Editor
Dr. David E. Matthews
Department of Statistics
University of Waterloo
Waterloo, Ontario/Canada N2L 3G1

Library of Congress Cataloging in Publication Data

Canadian Mathematical Congresses (Society)
Mathematics and the life sciences.

(Lecture notes in biomathematics ; 18)
Presented at the 15th biennial seminar of the Canadian Mathematical Congress, held at the Université de Sherbrooke.
Bibliography: p.
Includes index.
1. Mathematics--Addresses, essays, lectures.
2. Biomathematics--Addresses, essays, lectures.
I. Matthews, David E., 1948- II. Title. III. Series.
QA7.C25 1977 574'.01'51 77-11151

AMS Subject Classifications (1970): 57D45, 62P99, 92A05, 92A15

ISBN 3-540-08351-0 Springer-Verlag Berlin Heidelberg New York
ISBN 0-387-08351-0 Springer-Verlag New York Heidelberg Berlin

Printing and binding: Beltz Offsetdruck, Hemsbach/Bergstr.
2145/3140-543210

FOREWORD

For two weeks in August, 1975 more than 140 mathematicians and other scientists gathered at the Université de Sherbrooke. The occasion was the 15th Biennial Seminar of the Canadian Mathematical Congress, entitled *Mathematics and the Life Sciences*. Participants in this interdisciplinary gathering included researchers and graduate students in mathematics, seven different areas of biological science, physics, chemistry and medical science. Geographically, those present came from the United States and the United Kingdom as well as from academic departments and government agencies scattered across Canada.

In choosing this particular interdisciplinary topic the programme committee had two chief objectives. These were to promote Canadian research in mathematical problems of the life sciences, and to encourage co-operation and exchanges between mathematical scientists, biologists and medical researchers. To accomplish these objective the committee assembled a stimulating programme of lectures and talks. Six principal lecturers each delivered a series of five one-hour lectures in which various aspects of the interaction between mathematics and the life sciences were considered. In addition researchers working in the areas of health, population biology, physiology and development biology and disease processes were invited to give more than 25 hours of complementary talks.

The contributions of the principal lecturers concerned chiefly model building and verification in different areas of biology and the medical sciences. Many of the lectures were provocative and directed the attention of seminar participants to new areas of research. Some of the mathematical problems posed by the principal lecturers were solved at Sherbrooke. It was clear that the intended exchange between mathematical scientists and researchers in the life sciences was beginning to develop in the informal atmosphere encouraged by the Sherbrooke surroundings. This salutory effect was heightened by the many spontaneous group and individual discussions which occurred as the seminar progressed.

Initially, it was decided to reproduce the contributions of the principal lecturers (indicated by an asterisk in the table of contents), omitting those of Professor Hans Bremermann who decided to incorporate his contribution in a volume detailing his own research. The complementary lectures of Dr. Woodcock were subsequently chosen to replace those of Professor Bremermann. Regrettably, this selectivity does not do justice to the other excellent lectures that were given by many able seminar participants. It is the editor's hope that lectures which do not appear in the seminar proceedings will find deserving recognition elsewhere.

This volume, then, is only a sample of some of the seminar activities. It constitutes a permanent record of the stimulating atmosphere which infused the two weeks of lectures and discussions at Sherbrooke. It may also serve to acquaint the reader with some of the different perspectives from which mathematical and life science researchers, experimentalists and theoreticians alike, approach a host of contemporary problems.

Thanks are due to Honeywell Information Systems, the National Research Council and the Universities of Victoria and Waterloo; their generous financial support helped make the seminar possible. Thanks, also, to the proceedings contributors for their co-operation, to Professor W.F. Forbes for thoughtful editorial counsel, to the patient typists Annemarie Nittel,

Redina Caracaz and Julie Dembski, and to the publishers Springer-Verlag, New York. The readiness of the *Scandinavian Journal of Statistics* to permit the reprinting of a paper by Professor Barnard is gratefully acknowledged.

D.E. Matthews

Department of Statistics

University of Waterloo

TABLE OF CONTENTS

BARNARD, G.A.*	Problems of statistical inference in the life sciences	1
FINNEY, D.J.*	Bioassay	66
LEVINS, R.*	Quantitative analysis of complex systems	152
MAYNARD SMITH, J.*	Mathematical models in population biology	200
ROSEN, R.*	The generation and recognition of patterns in biological systems	222
WOODCOCK, A.E.R.	Catastrophe theory and the modelling of biological systems	342

* indicates principal lecturer

Problems of Statistical Inference in the Life Sciences

Professor G.A. Barnard

Department of Mathematics, University of Essex, Wivenhoe Park, Colchester, England

and

Department of Statistics, Faculty of Mathematics, University of Waterloo, Waterloo, Ontario, Canada

PROBLEMS OF STATISTICAL INFERENCE IN THE LIFE SCIENCES

Preface

In his R. A. Fisher Memorial Lecture the late Jimmie Savage told how, when he wanted to learn about statistics, his mentors advised him to work through Fisher's *Statistical Methods for Research Workers*, which he did. He went on to say that few people, he supposed, would think today of using that book as an introductory text. In my opinion this is a pity. My own introduction to statistics occurred early in 1933, when I carried out a survey among my schoolmates to ascertain (in statistical language not then understood by me) whether the correlation between a person's political stance and that of the newspaper he read was stronger if he first turned to the political comment than if he first turned to the sports pages. Naturally, I needed help, and this was provided by the late W. L. Stevens. When I remarked to him that I was interested in pursuing the subject further, but could find no suitable literature, he said I had better see his boss -- R. A. Fisher. When I had stated my difficulty to Fisher, he picked up a copy of 'Statistical Methods', and told me that if I read that work I would find in it many statements which called for proof. Being a mathematician, I ought to be able to work out the proofs for myself; and if I did so, then I would have learned mathematical statistics. The next time I met Fisher was nearly twenty years later (though we had been in correspondence), when to my amazement I learned he had appointed me as one of his vice-Presidents for the Royal Statistical Society. I was able to tell him that I had, just the week before, more or less completed the task he had set me nearly twenty years earlier. And I correspondingly felt I was beginning to understand mathematical statistics. There are, of course, other books -- for example, Bliss' *Statistics in Biology* -- which convey essentially the same message concerning the relationship of statistical methods to scientific research. And yet, to my knowledge there still does not exist a book as terse as Fisher's which succeeds in conveying the subtle relationship that must exist between statistical method and scientific judgement if statistics are not to be misused. Because of its terseness the book needs careful and repeated reading. Such reading would be greatly assisted -- and the unjustified reputation of excessive dogmatism would be rebutted -- if we had a 'variorum' edition, in which the evolution of Fisher's views could be traced from the first edition of 1925 to the final fourteenth edition of 1970. It would be clear from such an edition that, while ready to adopt a magisterial style on occasion, Fisher himself did not regard all the problems of statistical inference as having been solved. In particular, having clearly in mind the fact that any probability statement depends, not only on a careful specification of what is known, but also on a careful specification of what is not known, Fisher regarded the problem of specifying ignorance as one of the most difficult yet to be considered. Some indications of an approach to this problem are given in what follows.

Fisher had two principal reasons for adopting the 'Example' style which is characteristic of his book. Such a style exemplified the interaction of statistics with general sceintific considerations; it also kept the exposition simple. To further simplify matters Fisher made the valid assumption that almost all the continuous distributions to be encountered in biological contexts could, perhaps after a simple transformation, be taken as normal. I have departed from Fisher's principle of simplicity by attempting to show how the methods can be extended to the occasional case in which normal theory is not applicable. This not only extends the range of application of the theory but also serves to bring out the essential logic. Again, to keep matters simple, in his discussion of categorical data, Fisher treated the case of moderate sample sizes, in which the score function can be considered to be linear, with a slope virtually independent of the true parameter value. I have touched upon the issues that arise when such assumptions cannot be made.

One difficulty in giving a mathematical account of the theory of statistical estimation is that mathematical formulations of the problem to be solved are generally defective. In the account given here we propose to extend the usual scientific statement of an estimate in the form

$$\theta = t \pm s$$

(giving t as the estimate of θ, with standard error s); henceforth, we will specify the form of the distribution of the error of the estimate, writing

$$\theta = t \pm s \ (D)$$

to imply that the error, $t-\theta$, of the estimate has a scale factor s and a distribution of type D.

Notation and Conventions

Although most of the arguments used here extend to general probability distributions, discrete, continuous, or otherwise, we often tacitly assume we are dealing with the continuous case where the distribution is specified by a density function $\phi(x_1, x_2, \ldots, x_n)$, assumed continuous over the set where $\phi>0$. We will write $\underline{x}$ for the column vector

$$\begin{bmatrix} x_1 \\ x_2 \\ \cdot \\ \cdot \\ \cdot \\ x_n \end{bmatrix}$$

and abbreviate $\phi(x_1, x_2, \ldots, x_n)$ to $\phi(\underline{x})$, and $dx_1 dx_2 \ldots dx_n$ to $d\underline{x}$. The symbols $\underline{1}$ and $\underline{0}$ denote, respectively, column vectors of 1's and 0's and ' indicates transpose, so that $\underline{1}'$ represents a row of 1's . The lengths of these vectors are supposed chosen so as to make sense of the formulae in which they occur. We denote the identity matrix by I and represent the inverse of A by A^{-} .

When limits in an integral are unspecified, the integral is evaluated on the entire range of the variables.

To say that a density function ϕ is in standard form means usually that

$$\int \underline{x}\phi(\underline{x})\, d\underline{x} = \underline{0} \qquad \text{and} \qquad \int \underline{x}\,\underline{x}'\phi(\underline{x})\, d\underline{x} = I \ ,$$

but sometimes this convention is not followed. For example, it is more convenient to consider the function

$$u(x) = \begin{cases} 1, & |x| \le \frac{1}{2} \\ 0, & |x| > \frac{1}{2} \end{cases}$$

to be the standard form for the rectangular density. Similarly, since the variance of the Cauchy distribution does not exist we take

$$c(x) = 1/\pi(1 + x^2)$$

as the standard form for this distribution. We represent the standard form of the normal or Gaussian density by

$$n(\underline{x}) = (1/\sqrt{2\pi})^k \exp -\tfrac{1}{2}\underline{x}'\underline{x} .$$

In general, if ϕ is a density function, ϕ^* will represent its standard form. Beginning with $\phi(\underline{x})$ in standard form we can generate a *location parameter family* of distributions and densities by forming the family of densities $\phi(\underline{x} - \underline{\theta})$, where the vector $\underline{\theta}$ is then called the location parameter. If the standard form has been defined by the condition that the mean is zero, then the location parameter is the mean. In the case of the Cauchy distribution, however, the location parameter is the median. Similarly, if θ is a square matrix we can obtain the *scale parameter family* $\phi(\theta^{-}\underline{x})/|\theta|$, and when the standard form has been defined by the condition

$$\int \underline{x}\,\underline{x}'\phi(\underline{x})\,d\underline{x} = I$$

then $\theta'\theta$ is the variance-covariance matrix, often denoted by v . We obtain a *location and scale parameter family*, from a standard form $\phi(\underline{x})$, as $\phi(\theta_1^{-}(\underline{x} - \underline{\theta}_2))/|\theta_1|$, with θ_1 a matrix and $\underline{\theta}_2$ a vector. Conversely, if a density ϕ has location parameter $\underline{\theta}_2$ and scale parameter $\underline{\theta}_1$, then the standardised form ϕ^* will be given by

$$\phi^*(\underline{x}) = |\underline{\theta}_1|\phi(\underline{\theta}_1\ (\underline{x} + \underline{\theta}_2))$$

where $\underline{\theta}_1$ and $\underline{\theta}_2$ are suitably chosen.

The principal advantage of standardising by reference to the mean and variance is that, with this convention, if the variables $\underline{x}$ and $\underline{y}$ have their densities in standard form, and are statistically independent, then the linear compound $\underline{z} = \underline{a}\,\underline{x} + \underline{b}\,\underline{y}$ is also in standard form if and only if $\underline{a}\,\underline{a}' + \underline{b}\,\underline{b}' = I$. In addition, according to Chebyshev's inequality, if $\underline{x}$ is standardised in this way, $\Pr(|\underline{\lambda}'\underline{x}| \geq k\underline{\lambda}'\underline{\lambda}) \leq 1/k^2$. On the other hand, since finding the mean requires, in general, an integration, finding the mode may well be much easier.

We shall have to make considerable use of marginal and conditional distributions. If $\underline{x}$ is partitioned as $\begin{bmatrix} \underline{x}_1 \\ \underline{x}_2 \end{bmatrix}$, and we write the density of $\underline{x}$ as $\phi(\underline{x}_1,\underline{x}_2)$ then the marginal density of $\underline{x}_1$ is

$$\psi\,(\underline{x}_1) = \int \phi(\underline{x}_1,\underline{x}_2)\, d\underline{x}_2 \quad .$$

This density gives the probabilities of any proposition about $\underline{x}$ whose truth or falsehood depends only on $\underline{x}_1$ and is sometimes loosely spoken of as specifying the density for $\underline{x}_1$ *regardless* of $\underline{x}_2$, but this is not correct. To say *regardless* would be to suggest that no account need be taken of any present or future knowledge of the value, or possible values of $\underline{x}_2$. But since statements of probability must be based, so far as possible, on all available relevant knowledge, if it came to be known, for instance, that $\underline{x}_2$ belonged to a set S , then the probability density for $\underline{x}_1$ would not be given by the marginal density $\psi(\underline{x}_1)$, but by

$$K() \int_S \phi(\underline{x}_1,\underline{x}_2) d\underline{x}_2 \quad .$$

In particular, use of the marginal density for $\underline{x}_1$ *requires, in principle, that* $\underline{x}_2$ *be known to follow its own marginal distribution.* In the above expression, $K()$ is used to denote a constant (i.e. a quantity not varying with $\underline{x}_1$) chosen to normalise the density of $\underline{x}_1$ to make it integrate to 1. By using the special symbol $K()$ we allow for possible changes in the constant to incorporate further factors not involving the variable to which the density refers.

If $\underline{x}_2 = \underline{a}$, but $\underline{x}_1$ remains unknown in an experiment governed by the joint density $\phi(\underline{x}_1,\underline{x}_2)$, then the relevant density for $\underline{x}_1$ becomes the conditional density, $K()\phi(\underline{x}_1,\underline{a})$, obtained by replacing $\underline{x}_2$ by its known value, and renormalising. The value of $K()$ here will in fact be the reciprocal of the marginal density of $\underline{x}_2$, evaluated at $\underline{x}_2 = \underline{a}$.

For mean values we adopt the convenient notation of physicists, denoting the mean value of x by $\langle x \rangle$. When it is necessary to specify the value θ of a parameter on which this mean value depends we write $\langle x \rangle_\theta$.

We assume familiarity with Bayes' Theorem which, in its application to propositions A and B, derives from the two ways of writing $Pr(A \& B)$; thus

$$Pr(A\&B) = Pr(A)Pr(B:A) = Pr(B)Pr(A:B)$$

from which, if A is taken as fixed and known,

$$Pr(B:A) \propto Pr(A:B)Pr(B) .$$

In terms of density functions for continuous distributions, Bayes' theorem gives the posterior density for θ, given observations $\underline{x}$, and prior density $\pi(\theta)$, as

$$P(\theta) = K()\phi(\underline{x}:\theta)\pi(\theta)$$

where $\phi(\underline{x}:\theta)$ is the density function for the observations $\underline{x}$, involving the unknown θ. When regarded as a function of θ for given $\underline{x}$, $\phi(\underline{x}:\theta)$ is called the *likelihood function* for θ given $\underline{x}$. Because of $K()$, any multiple of ϕ by a quantity which is not a function of θ gives an equivalent likelihood function.

If firm prior information in terms of a given $\pi(\theta)$ is available, the problem of estimation becomes trivial. The general estimate is best specified by specifying the posterior density $P(\theta)$. By means of a linear standardising transformation $\theta \to \theta^* = (\theta - \bar{\theta})/s$ where

$$\bar{\theta} = \int \theta P(\theta) d\theta \; , \qquad \text{and} \qquad s^2 = \int (\theta - \bar{\theta})^2 P(\theta) d\theta$$

we find the density function of θ^* as P^* , with zero mean and unit variance, and then we can write

$$\theta = \bar{\theta} + s(P^*)$$

where (P^*) is taken to mean a random variable having density P^* .

Questions are marked as Qn; it should be understood that not all these questions questions have a well-determined answer.

Q0: Find the values t to take for θ to minimise the mean losses, when the losses are taken as (a) $(t-\theta)^2$, (b) $(t-\theta)^4$, (c) $t-\theta$ when $t \geq \theta$ and $2(\theta-t)$ when $t<\theta$, (d) $(t-\theta)^2$ when $|t-\theta|<M$ and M^2 when $|t-\theta|>M$.

Two Aspects of Estimation

The statistical theory of errors arose when it was noticed that what were essentially repeated measurements on the same set-up gave discordant results. The discordances were attributed to errors of observation which were regarded as having a definite probability distribution. Two problems arose: (i) reducing the discordant set of observations to a single observation, or to a concordant set; (ii) specifying the error distribution to which the reduced observation was subject. A further problem sometimes arose -- for example in the construction of maps -- that of determining a 'best' value to be used for a quantity that had been measured subject to error.

Subsequently, the theory of statistical estimation was extended to cases where the processes underlying natural phenomena were regarded as probabilistic in character, with the probabilities being functions of unknown parameters (such as the recombination fraction in genetics). Methods were required for combining the evidence from different types of data bearing on the same parameter -- for example, from different types of crossings in genetics -- for assessing the consistency or otherwise of the different sets, and for expressing the resultant information in the form of a 'quasi-measurement', that is, a single number, calculated from the observations, which could be regarded as the result of a measurement of the quantity in question.

In the course of the XIXth century it came to be accepted that errors of measurement in astronomy, and in most other fields, were at least approximately normally distributed; thus, the result, y, of a single measurement on a quantity θ could be expressed as

$$y = \theta + b + e$$

where b was a 'bias' inherent in the method of measurement, and e was a random error normally distributed about zero with root mean square value σ, called the 'standard error'. Such a result would be expressed as the 'estimate'

$$\theta = (y - b) \pm \sigma$$

where the bias would have been corrected by subtraction from y and the conventional $\pm$ sign was used to indicate the scale factor of the error distribution. As a matter of fact, the use of the r.m.s. value σ became common only in the XXth century; for most of the XIXth century the 'probable error', $0.6745\,\sigma$, was quoted instead. Since half the distribution in the standard normal curve lies between the limits $\pm\,0.6745$ it was argued that the true value θ was as likely as not to lie between the limits $(y-b) \pm 0.6745\sigma$ A rule of thumb also developed which regarded any discrepancy between a theoretical value and a measured value that exceeded three times the probable error as suspicious. The same rule rejected any discrepancy exceeding four times the probable error as incredible. Thus when Galle discovered a planet near the position predicted by Leverrier and Adams for a planet which would account for the discrepancies in the orbit of Uranus, the American astronomer Benjamin Pierce declared that Galle's planet could not be the one predicted, because the distance between the predicted position and that observed was more than four times the probable error. The discrepancy was later attributed to a wrong value, taken from Bode's law, for the distance of Neptune from the Sun. Nowadays the 'suspicious' limit corresponds to a tail probability of 0.05 , while the 'rejection' limit is associated with a probability of 0.01 . These conventional limits, popularised by R. A. Fisher, were chosen by him to correspond roughly to the older limits.

When Fisher first published his *Statistical Methods for Research Workers* he wished to include tables of the most frequently used statistical distributions, but encountered difficulties because the copyright in most of these was held by the Biometrika Trust, controlled by Karl Pearson. Pearson always pursued a strong line on copyright in order to help secure the financial success of the Biometrika Trust, and in view of the acrimonious personal relations between him and Fisher at the time it was hardly to be expected that Pearson would go out of his way to make a special exception; for similar reasons Fisher would not have found it easy to ask for special consideration. And so, in the tables for his book, Fisher gave the deviate as a function of the tail area probability, rather than the probability as a function of the deviate. It must be said, however, that both men were ultimately devoted to the advancement of science, and neither would, I think, have consciously allowed personal considerations to stand in the way of scientific advance. As we shall see, Fisher had reasons for thinking there might be advantages in the form of tabulation he adopted. It is doubtful, however, whether he fully appreciated the way in which the 5% and 1% values would become so rigidly enshrined in statistical practice in so many areas; had he done so, he might have reconsidered his decision.

To summarize, then, the most common form for presenting an 'estimate' is

$$\theta = t \pm s \tag{1}$$

which is understood to mean that the information available about θ is equivalent to that which would be obtained from a direct, unbiased measurement giving the value t, with an instrument having standard error s.

The problem of statistical estimation primarily consists of reducing complex sets of data to the canonical form (1), or something similar, in the most efficient manner possible. In addition, there sometimes arises the problem of deriving a 'best' value, to be used in some specified context without a probable error. An early example of this last type was considered by Gauss in connection with his duties as head of the German Geodetic Survey - in making a map he had to choose a single figure to take as the distance from the centre of Berlin to the centre of Potsdam since any element of uncertainty in the value taken could not be conveniently represented. As another example, when giving a patient a dose of a drug you must fix a single figure which allows as best you can for your estimate of his tolerance for the drug, and for the variation in its effect. These are decision problems, and will inevitably involve gains and losses, specific to each problem, which must be taken into account. We should note that because the (D) part of our estimate does not necessarily refer to a symmetric distribution, the $\pm$ sign is inappropriate, if taken in its ordinary sense. It should perhaps be replaced by "-" , since our interpretation is that $(t-\theta)/s$ has density D . But if read as "more or less" or "subject to error" the $\pm$ sign can usefully be retained.

To distinguish between the problem of reduction to canonical form and the decision type of problem we may refer to the former as the *general estimation problem*, and to the latter as the *specific parametric decision problem*. The words 'general' and 'specific' serve to indicate that the second type of problem involves a specification of losses and gains, and perhaps of other features not required in the general estimation problem.

A Simple Illustration

Suppose I have in my mind a large whole number N -- for instance, the number of my railway ticket. Without telling you the value of N , and without showing you the cards, I cut a standard, well-shuffled deck and then tell you the value of

$$X = N + E$$

where E is the numerical value of the face card, counting $A = 1$ and $K = 13$. Given that $X = 3215$, what do you know about N ?

So far as the general estimation problem is concerned, since the error E has mean value $+7$ you could write

$$N = (3215-7) \pm 6 \quad ,$$

provided you indicated that 6 was not the r.m.s of a Gaussian error distribution but the semirange of a rectangular distribution. This might be done by putting (R) after the 6; thus

$$N = 3208 \pm 6(R) \quad .$$

Then, if a specific decision problem arose concerning N, a problem involving a large penalty for over-estimation and a small penalty for moderate underestimation (say, not more than 13 units), the solution would clearly be to take 3202 as the value. However, if the penalty for error was proportional to the square of the error, or to any higher even power, a strong case for taking N to be 3208 could be advanced.

Q1: Why? What if it were known that railway ticket numbers were always odd? What if it were known that they were always four digit numbers beginning with 3 ?

Q2: What value would you take for N in a specific decision problem if the penalty P for a mistake $M = \tilde{N} - N$ (where $\tilde{N}$ is the number you take, and N is the true value) is

(a) $P = 0$ if $|M| \leq 6$, $P = 1$ otherwise;

(b) $P = M$ if $M \geq 0$, and $P = -2M$ if $M<0$.

Now suppose that I cut the cards again and tell you, if the face card this time has value E' , that

$$X' = N + E' = 3213.$$

Obviously we could give the estimate

$$N = 3206 \pm 6(R) .$$

How can we combine our two estimates? Common sense tells us that N must lie between 3202 and 3212 , so that we are tempted to write

$$N = 3207 \pm 5(R) .$$

But can we justify the (R) ? It amounts to saying that our knowledge of N is equivalent to that which we would have had on being told

$$Z = N + F = 3213$$

where F is the face value of a card cut from a deck from which all the picture cards except the jacks have been removed. Can we deduce this from the two pieces of information provided?

From the first observation we know that $X = 3215$ while $E = X-N$ is uniformly distributed from 1 up to 13 . From the second we know that $X' =$ 3213 , while $E' = X'-N$ is uniformly distributed from 1 up to 13 , independently of E . Now the joint distribution of (E,E') can be specified in terms of (Q,Q') where

$$Q = E$$
$$Q' = E'-E ,$$

since this transformation is 1 to 1 ; to every proposition about (E,E') there corresponds a logically equivalent proposition about (Q,Q') . To describe the information in terms of the values of (X,X') and the distribution of (E,E') is logically equivalent to describing it in terms of $(X, X'-X)$ and (Q,Q') . But

$$Q' = E'-E = (X'-N) - (X-N) = X'-X$$

has a known value -2 ; hence, in the new description the only unknown quantity is $Q = X-N$, and the probability distribution of this should clearly be calculated given the known value of Q' . Given that $Q' = -2$, E can range only from 3 up to 13 , and each of the possible values has equal probability. Thus we deduce that our information about N amounts to an observation $X = 3215$, with error Q uniformly distributed between $3,4,\ldots,13$. Correcting for bias, as before, we would express this as

$$N = (3215-8) \pm 5(R) = 3207 \pm 5\ (R) .$$

Now suppose I cut my original card deck a third time, and tell you

$$X'' = N + E'' = 3204 .$$

Defining
$$Q'' = E''-Q = X''-X = 3204 - 3215 = -11$$

and conditioning the distribution of Q , given $Q'' = -11$, we find that Q can now take only the values 12 and 13 , and conditionally these are equally probable. Therefore we deduce that

$$N = (3215 - 12.5) \pm 0.5\ (R)$$
$$= 3202.5 \pm 0.5\ (R) .$$

The three observations taken together reduce the possible values of N to

3202 and 3203, each being equally plausible so far as the data alone are concerned.

Q3: What is the probability, in this game, that you would know the value of N exactly after 3 cuts (at most)? Hint: For this to happen, one of the three cuts must have given 1, while another must have given 13. Thus the possibilities are (1,13,A), (1,A',13), (A",1,13), (13,1,A), (13,A*,1), (A",13,1), where A denotes any number 1 to 13, A' omits 13, A" omits 1 and 13, and A* omits 1.

Ans. $72/13^3$, or about 1/30.

Q4: How many cuts, on the average, before N is known for certain?

Ans. 19.5

---

* Is 'Equally plausible' Equivalent to 'Equally probable'?

It is sometimes argued that in order for two events to be equally probable, over a long series of independent trials there should be no rule which would guarantee us a win betting at even odds, even if we have, on each occasion, the choice of betting either way or not at all. But consider the following rule: with reduced observation

$$\hat{N} = N + G$$

with G equally probably +.5 or -.5, bet on $N = \hat{N} + .5$ if $\hat{N} \leq 3212.5$ and on $N = \hat{N} - .5$ if $\hat{N} \geq 3212.5$. With this rule, if in fact $N = 3212$, we win every time, since the estimate must be either

---

* starred sections may be omitted on first reading.

3211.5 or 3212.5, and in either case we bet on 3212. But if, for instance, the true value is 3211, half the time the estimate will be 3211.5, in which case we lose, while half the time the estimate will be 3210.5, in which case we bet on 3211 and win. Similarly in all other cases, whenever the true value is $\neq$ 3212, we win half the time and lose half the time and so, since we are betting at even odds, in these cases we on balance neither win nor lose. But whenever $N = 3212$ we win every time.

The fact that we can produce a rule which guarantees a win in some cases, while on the average losing nothing in other cases, is taken by many people to imply that we cannot regard the two values, estimate +0.5 and estimate −0.5 as equally *probable*. Although, before the value of the estimate was known, the probabilities of errors +0.5 and −0.5 were equal, we can classify the observed estimates into a 'high' set (3212.5 or more) and a 'low' set (3211.5 or less); and the chance of a positive error in the high set is somewhat larger than the chance of a positive error in the low set, and conversely. Thus after the value of the estimate is known, we can no longer regard the error distribution as the same as it was before. Others point to the fact that, given any positive ε, however small, we could, without violating any of the given conditions, have the probability of positive error in the high set less than $0.5 + \varepsilon$. That is, this probability cannot be bounded away from 0.5.

In any given situation, so long as we know how to interpret an observed value, say 3211.5, which is subject to equally probable errors ±0.5, we need not enter into these questions further, at this time. We shall see that most questions of the sort that arise in scientific contexts can be answered without settling these issues. Problems sometimes arise in connection with specific decisions, however.*

General Method of Estimating a Location Parameter

A general problem, of which the railway ticket illustration is a special case, can be formulated as follows: We are given n independent observations x_i $(i=1,2,\ldots\ldots,n)$ each of which has probability function $\phi_i(x_i-\theta)$ involving the unknown location parameter θ . The ϕ_i are known. We wish to estimate θ ; that is, we wish to summarise the information about θ in the form

$$\theta \quad = \quad t \pm s(D) \tag{2}$$

where this expression is taken to mean that we have the equivalent of an observation t from a distribution whose shape is specified by D and which has location parameter θ and scale parameter s . An example considered in many elementary statistics texts is that where, for all i , ϕ_i is the standard normal density

$$\phi_i(u) \quad = \quad (1\sqrt{2\pi}) \exp -u^2/2 \quad ; \tag{3}$$

in this case we have

$$\theta \quad = \quad \bar{x} \pm (1/\sqrt{n}) \; (N) \tag{4}$$

where N denotes the normal density, in standard form. We should however stress that for our purposes ϕ_i can be quite arbitrary -- normal, rectangular, Cauchy, or whatever. The more realistic case where the scale parameter as well as the location parameter is unknown will be considered later.

---

* starred sections may be omitted on first reading

Following the pattern of the railway ticket argument with more general terminology, we introduce the term *pivotal* to denote a function of the unknown θ and of the observation(s) whose distribution is known completely. Thus $x_i - \theta$, for each i , is a pivotal, since its probability function is the known ϕ_i . The vector $\underline{p} = \underline{x} - \theta.\underline{1}$ is also a pivotal, since its density function is

$$\psi(\underline{p}) = \prod_i \phi_i(p_i) \tag{5}$$

in view of the fact that the observations are independent. We can then state our problem formally as follows:

In the given experiment, $\underline{p} = (\underline{x} - \theta.\underline{1})$ has density $\psi(\underline{p})$. The observed result is $\underline{x} = \underline{x}_o$. What can be deduced about θ ?

Now the $n \times n$ matrix

$$T = \begin{bmatrix} 1 & \underline{0}' \\ \underline{1} & -I \end{bmatrix}$$

is invertible (in fact $T^- = T$) , so the transformation

$$\underline{p} \to \underline{q} = T\,\underline{p}$$

is one to one, and its Jacobian is 1 . Therefore, to say that $\underline{p}$ has density $\psi(\underline{p})$ is logically equivalent to saying that $\underline{q}$ has density $\psi(T\underline{q})$. And to say that the observed result is $\underline{x} = \underline{x}_o$ is equivalent to saying that the observed result is $\underline{y} = \underline{y}_o$, where $\underline{y} = T\,\underline{x}$ and $\underline{y}_o = T\,\underline{x}_o$. Thus we can restate our problem in the logically equivalent form:

> In the given experiment, $\underline{q} = \underline{y} - \theta \begin{bmatrix} 1 \\ \underline{0} \end{bmatrix}$ has density $\psi(T\,\underline{q})$.
>
> The observed result is $\underline{y} = \underline{y}_o$.
>
> What can be deduced about θ ?

We have here used the fact that $T\,\underline{1} = \begin{bmatrix} 1 \\ \underline{0} \end{bmatrix}$.

But in this transformed version of the problem the only component of $\underline{q}$ whose value has not been observed is the first, viz. $y_1 - \theta$. The second component, for instance, is $y_1 - y_2$, whose value has been observed as $y_{1o} - y_{2o}$. Since all components of $\underline{q}$ but the first are known, the relevant density for this first component is the conditional density, obtained by substituting $q_2 = y_{2o}, \ldots, q_n = y_{no}$ in $\psi(T\,\underline{q})$, viz.

$$\xi\,(q_1) = \psi(q_1, q_1 - y_{2o}, \ldots, q_1 - y_{no}).K() \quad .$$

Therefore, by applying the principle of calculating probabilities on all the relevant available information, we reduce the problem to:

> In the given experiment, $q_1 = y_1 - \theta$ has density $\xi(q_1)$.
>
> The observed result is $y_1 = y_{1o}$.
>
> What can be deduced about θ ?

Now suppose that the standardised form of the density ξ is ξ^* , and that ξ has location parameter b and scale parameter s . Since to say that q_1 has density $\xi(q_1)$ is logically equivalent to saying that $q_1 - b$ has density $\xi((q_1-b)+b)$, and to say that y_1 is observed to equal y_{1o} is logically equivalent to saying that y_1-b is observed to equal $y_{1o}-b$, from the interpretation of an estimate given in equation (2) , we can deduce for θ the estimate

$$\theta = (y_1-b) \pm s(\xi^*) \quad . \tag{6}$$

That is, our experimental information about θ is equivalent to that from a direct measurement of θ by an instrument whose error distribution has the shape specified by ξ^* , and the scale factor s , when the value of the reading is (y_1-b) .

Examples

(1) If $\phi_i(p_i) = n(p_i)$ then

$$\psi(\underline{p}) = K() \exp(-\tfrac{1}{2}\underline{p}'\underline{p})$$

so that

$$\xi(q_1) = K() \exp(-\tfrac{1}{2}q_1^2) \cdot \exp\{-\tfrac{1}{2}\sum_{i=2}^{n} (q_1-y_i)^2\}$$

$$= K() \exp(-\tfrac{1}{2}nq_1^2) \cdot \exp(q_1 \sum_{i=2}^{n} y_i)$$

absorbing $\exp(-\tfrac{1}{2}\Sigma\, y_i^2)$ into $K()$. Now

$$\sum_{i=2}^{n} y_i = \sum_{i=1}^{n} (x_1-x_i) = n(x_1-\bar{x})$$

so, changing $K()$ again,

$$\xi(q_1) = K() \exp[-\tfrac{1}{2}n\{q_1 - (x_1-\bar{x})^2\}]$$

so that ξ^* is the standard normal density, and the location and scale parameters of ξ are $(x_1-\bar{x})$ and $1/\sqrt{n}$, respectively. We thus arrive at the estimate

$$\theta = x_1-(x_1-\bar{x}) \pm (1/\sqrt{n})\ (N)$$

$$= \bar{x} \pm (1/\sqrt{n})\ (N) ,$$

i.e. the sample mean, with normally distributed error and standard error $1/\sqrt{n}$. This, of course, accords with usual practice.

(2) If $\phi_i\ (p_i) = u(p_i)$ then

$$\psi(\underline{p}) = K() \prod_{i=1}^{n} u(p_i)$$

and $\prod_{i=1}^{n} u(p_i)$ is 1 or 0 according to whether or not $|p_i| \le \frac{1}{2}$ for all i . It can be seen directly from this that $\xi^* = u$, and a little analysis shows that the estimate is

$$\theta = m \pm w(R)$$

where $m = \frac{1}{2}\{\mathrm{Max}(x_i) + \mathrm{Min}(x_i)\}$, and $w = \mathrm{Max}(x_i) - \mathrm{Min}(x_i)$. The symbol (R) indicates that the error distribution is rectangular, for which the standardised density is u .

(3) If $\phi_i(p_i) = c(p_i)$ then

$$\psi(\underline{p}) = K()/\{\prod_{i=1}^{n} (1 + p_i^2)\} .$$

and $$\xi(q_1) = K()/[(1+q_1^2)\prod_{i=2}^{n}\{1+(q_1-y_i)^2\}]$$

The mean and variance of $\xi(q_1)$ exist for $n>1$; the mean is

$$b = \frac{\int wdw/[(1+w^2)\prod_{i=2}^{n}\{1+(w-y_i)^2\}]}{\int dw/[(1+w^2)\prod_{i=2}^{n}\{1+(w-y_i)^2\}]}$$

and for the variance s^2 we have a similar expression, with $(w-b)^2dw$ in the numerator instead of wdw . We then derive the estimate

$$\theta = t \pm s\ (\xi^*)$$

where

$$t = \frac{\int wdw/[\prod_{i=1}^{n}\{1+(x_i-w)^2\}]}{\int dw/[\prod_{i=1}^{n}\{1+(x_i-w)^2\}]}$$

which is the so-called Pitman estimate for this case. The shape ξ^* of the error distribution depends on the y_i , and it can be multimodal. Surprisingly often ξ^* is well approximated by the normal distribution. However it can fail to be symmetrical, in which case previous comments concerning the interpretation of "$\pm$" should be borne in mind.

What can be Deduced from an Estimate?

It was argued by Fisher (to whom the essence of the above argument is due) that in cases such as those considered above, where all of the experimental data about θ could be summed up in the estimate

$$\theta = t \pm s(D) ,$$

it was permissible, *in the absence of any other knowledge about* θ , to regard θ as itself a random variable having a distribution of shape specified by D , with location parameter t and scale parameter s . Others have found difficulty in accepting Fisher's view, owing to problems of the sort discussed in the section 'Is equally plausible equivalent to equally probable' , and some other types of difficulty. The present writer accepts Fisher's view -- noting that much depends on a careful specification of what is meant by absence of knowledge -- but it is important to stress here that from the practical point of view very little is to be gained from pursuing the controversy. We shall see later on that some Fisherian arguments which appear to depend on the fiducial argument (as he called it) in fact do not require its acceptance, and many practical applications of estimates of the type considered will be the same, irrespective of whether the argument is accepted or not. Perhaps one of the most important points to be stressed is that an estimate of the form (2) can be combined in a natural way with other probabilistic information about θ .

Suppose, for example, that θ denotes the viscosity of a lubricant of a type which is relatively well known, so that the lubricant in question may for this purpose be regarded as randomly chosen from this type. If the viscosities of the type in question are known to be reasonably uniformly distributed between the limits 10 - 30 units, and we have an estimate

$$\theta = 15 \pm 2(N) ,$$

this corresponds to an observation $t = 15$, normally distributed about θ

so that $(t-sa, t+sb)$ is a 90% confidence interval for θ. Correspondingly, if a theoretical value θ_o for θ is proposed, we can test its consistency with the data yielding our estimate by evaluating the tail area

$$\int_{|t|>t_o} \xi^*(t)\,dt$$

where $t_o = (t-\theta_o)/s$. If this tail area is small we shall be disinclined to accept the hypothesis $\theta = \theta_o$. Finally, if another experiment yields the estimate

$$\theta = t' \pm s'(\xi'^*)$$

then by means of an argument essentially similar to that which led to $t \pm s(\xi)$ in the first place, taking $n = 2$, $x_1 = t$, $x_2 = t'$, $p_1 = t-\theta$, $p_2 = t'-\theta$, $\phi_1(p_1) = (1/s)\xi^*(p_1/s)$, $\phi_2(p_2) = (1/s')\xi'^*(p_2/s')$, the conditional distribution of $q_1 = t-\theta$, given the known value of $q_2 = t-t' = d$, say, is given by the density

$$\xi''(q_1) = K()\xi^*(q_1/s)\xi'^*((q_1-d)/s') .$$

This leads to the combined estimate

$$\theta = t'' \pm s''(\xi''^*)$$

where

$$t'' = \frac{\int z\, \xi^*\left(\frac{t-z}{s}\right) \xi'^*\left(\frac{t'-z}{s'}\right) dz}{\int \xi^*\left(\frac{t-z}{s}\right) \xi'^*\left(\frac{t'-z}{s'}\right) dz} .$$

with standard deviation 2, i.e. with probability function

$$\tfrac{1}{2}n(\tfrac{1}{2}(t-\theta)) \quad .$$

Inserting the value $t = 15$ gives us a likelihood function by which the uniform prior density $u((\theta - 20)/20)/20$ can be multipled to obtain, in accordance with Bayes' Theorem, a posterior density

$$K() \; n(\tfrac{1}{2}(15-\theta)).u((\theta-20)/20)$$

and this, apart from the cutting off of the lower tail below $\theta = 10$ corresponds very closely to a normal distribution centred on 15, with standard deviation 2. It should be noted that exactly the same posterior distribution would have been obtained by applying Bayes' Theorem to the original data of the experiment. Because of the equivalence of the information in the estimate and the information in the original data, the likelihood functions are the same.

If, on the other hand, external information of a probabilistic kind is lacking, we can use the estimate to derive confidence limits for θ. If a and b are limits such that

$$\int_a^b \xi^*(t)dt = 0.90 \quad ,$$

then the meaning of our estimate tells us that

$$\Pr\{a < (t-\theta)/s < b\} = 0.9$$

so that in a long run of repeated experiments, we shall have

$$a < (t-\theta)/s < b$$

true in 90% of them. But (7) is true if and only if

$$t - sa < \theta < t + sb$$

Q5: Verify that the result of combining the estimates

$$\theta = t \pm s(N)$$

and

$$\theta = t' \pm s'(N)$$

is

$$\theta = t'' \pm s''(N)$$

where

$$t'' = (s'^2 t + s^2 t')/(s'^2 + s^2)$$

and

$$s''^2 = (s's)^2/(s'^2 + s^2) \quad .$$

It may be useful at this point to summarise the preceding results in the form of a theorem.

"Theorem"

If the observations x_i $(i=1,2,\ldots,n)$ are related to the scalar location parameter θ by the known joint density ϕ ,

$$\phi = \phi(x_1-\theta, x_2-\theta, \ldots, x_n-\theta)$$

then the solution to the general estimation problem is

$$\theta = t \pm s(L^*(-e))$$

where L^* is the standardised form of

$$L(\theta) = \frac{\phi(x_1-\theta, x_2-\theta, \ldots, x_n-\theta)}{\int \phi(x_1-\theta, x_2-\theta, \ldots, x_n-\theta)\,d\theta}$$

and $$t = \int zL(z)dz \quad ,$$

while $$s^2 = \int (z-t)^2 L(z)dz$$

and $$L^*(-e) = sL(s(t-e)) \quad .$$

"Proof"

We have put theorem and proof between quotation marks to indicate that the argument which follows might be called 'paramathematical', in so far as it appeals to our understanding of imprecise terms such as 'experiment', 'information', 'known', etc.

If $p_i = x_i - \theta$, then to know that the p_i's have joint density $\phi(\underline{p})$ in an experiment, and that in such an experiment the observation $\underline{x}$ takes the values $\underline{x}_o$ is equivalent to knowing that in the same experiment, $\underline{q} = T\underline{p}$ has joint density $\phi(T\underline{q})$ and the transformed observations $\underline{y} = T\underline{x}$ take the values $\underline{y}_o = T\underline{x}_o$, where

$$T = T^{-} = \begin{bmatrix} 1 & \underline{0}' \\ \underline{1} & -I \end{bmatrix}.$$

If the vector $\underline{y}$ of transformed observations can be partitioned as $\underline{y} = \begin{bmatrix} y_1 \\ \underline{y}_2 \end{bmatrix}$ in such a way that the marginal density of $\underline{y}_2$ does not involve any unknown parameter, then the same inference about any unknown parameters may be drawn from observing $\underline{y} = \begin{bmatrix} y_{1o} \\ y_{2o} \end{bmatrix}$ as may be drawn from an experiment in which the observation is y_1 and its density function is the joint density of $\underline{y}$, conditioned on $\underline{y}_2 = \underline{y}_{2o}$. Now the marginal density of $(y_2,\ldots,y_n)$ is that of $(p_1-p_2,\ldots,p_1-p_n)$, viz.

$$\int \phi(p_1,p_1-p_2,\ldots,p_1-p_n)dp_1$$

which does not involve any unknown parameter. Therefore the same inference may be drawn from the observations $\underline{y} = \underline{y}_o$ as may be drawn from an experiment in which the observation is $y_1 = y_{1o}$ with the (conditional) density for y_1 given by

$$\xi(y_1-\theta) = K()\phi(y_1-\theta, y_1-\theta-y_{2o}, \ldots, y_1-\theta-y_{no}) .$$

Now if $z_o = y_o - b$ the same inference may be drawn from an observation $z = z_o$ when z has density $\xi(z+b)$ as may be drawn from an observation $y = y_o$ when y has density $\xi(y)$. But to know that $y = y_o$ is equivalent to knowing that $z = y - b$ is equal to $z_o = y_o - b$, and the density of z will be $\xi(z+b)$ if that of y is $\xi(y)$. Taking $y = y_1-\theta$ and $b = y_1-t$ with t as defined in the "theorem", we deduce that our information about θ is equivalent to that in an observation t from an experiment in which $t-\theta$ has density $L(t-\theta)$. The remainder of the "theorem" involves the standardising of the density function L , and assumes that L has a finite variance. Should this not be the case, another method of standardisation would have to be used, raising inessential complications.

Discussion

The notion that a 1-1 transformation of the observations in an experiment, along with the corresponding transformation of their probability function, does not alter the inference to be drawn will surely be generally acceptable. The less generally acceptable part of our "proof" involves the conditioning argument, that if the observations are $(y_1, \underline{y}_2)$ and the marginal density of $\underline{y}_2$ does not involve the unknown θ , then the experiment is equivalent to one in which y_1 was observed, having the conditional density given $\underline{y}_2 = \underline{y}_{2o}$. It has sometimes been argued that such conditioning results in loss of information, but as shown in the appended note "Conditional Inference is not Inefficient", this impression arises from a misunderstanding.

The disturbing aspect of the conditional argument is that cases can arise in which two transformations, $\underline{y} = T\underline{x}$, $\underline{z} = T'\underline{x}$ exist, such that $\underline{y}$ can be partitioned as $(\underline{y}_1, \underline{y}_2)$, with $\underline{y}_2$ distributed without dependence on θ, and $\underline{z}$ can be similarly partitioned, but the experiments corresponding to the observation of $\underline{y}_1$ and of $\underline{z}_1$ do not appear to be equivalent. This possibility arises from the fact that $\underline{y}_2$ may have a density not involving θ, and $\underline{z}_2$ also, but the joint density of $(\underline{y}_2, \underline{z}_2)$ may involve θ. A simple example is that of a pair of variables each having a standard normal density, but having correlation θ. Thus it seems that if we accept the idea of conditioning, we must accept the idea that the inferences to be drawn from an experiment may not be unique. But this is, in fact, an idea which we must accept anyway, unless we wish to restrict the definition of an experiment in what would seem to be an artificial way. For it is possible that an experiment may consist in the observation of two variables, y and z, each having a density depending on θ, for example

$$\phi_1(y) = n(y-\theta), \qquad \phi_2(z) = u((z-\theta)/4)/4$$

but we may have no idea of the joint density of (y,z). For purposes of inference we may use the value of y or the value of z, but we cannot put the two together. And there is no obvious way of choosing whether to base our inference on y or on z. If we were concerned to establish relatively wide limits beyond which we could be sure θ would not fall, we would evidently use z. But if we were prepared to run some risk of error, in order to achieve limits closer than ± 2, we would use y. It is easy to see that the relation between y and z can vary from complete statistical independence (and, of course, if we knew this was so we could infer from y and z jointly) to complete functional dependence, and this could take various forms. Ordinarily, in practice we would make a series of alternative assumptions, exhausting the range of reasonable possibilities, and make inferences subject to these assumptions in turn.

In case it should be thought that these possibilities are purely theoretical, one may cite the situation which in fact occurred within the present writer's experience. He was interested in the accident rate, per passenger-mile, of a car. He had data giving the accident rate as a function of the engine horse power, and data giving the accident rate for various makes of car. But no data were available giving accident rates as a function of both make and engine capacity. Thus, faced with a car of given engine capacity and given make, there was a choice of which datum to use to predict the accident rate. In cases such as this, if one datum clearly gives more information about accident rate than the other, then this will clearly be the one to use. But situations can readily be envisaged (as with the observations y and z above) when there is no general, clear cut preference.

In the location parameter case, however, it can be shown that the conditional inference is essentially unique, and these ambiguities do not arise.

It should perhaps be remarked that in the "Theorem" we have assumed that the density (in z)

$$K()\ \phi(x_1-z, x_2-z, \ldots, x_n-z)$$

possesses finite mean and variance. Neither of these conditions need be satisfied in all cases. But if, for example, we are concerned with independent and identically distributed observations with common density ϕ, provided that

$|\phi(x)| = o(1/x)$ as $|x| \to \infty$ then $z\phi(x_1-z)$ will be bounded, and so $\int z\phi(x_1-z)\phi(x_2-z)dz$ will converge. Similarly, provided $n > 3$, with the same condition on ϕ, the variance integral will converge. As remarked earlier, should the variance fail to exist, then some other method of standardising the density function would have to be used.

Relationship to Maximum Likelihood

The reader who has heard of the method of maximum likelihood will by this time be wondering what is the connection, if any, between our approach and that very useful method. The method consists in writing down the joint density of the observations evaluated at the observed values, as a function of the unknown parameter, viz.

$$L(\theta) = \phi(x_1-\theta, x_2-\theta, \ldots, x_n-\theta) \quad .$$

The likelihood function, $L(\theta)$, is defined only up to a multiplicative constant (i.e. a factor not dependent on θ), so that in order to make the representations of the likelihood unique, L must be standardised in some way. In the location parameter case a convenient standardisation is to require L to integrate to 1 over the range of θ, but perhaps the most widely used standardisation is to require that $\sup_{\theta} L(\theta)$ should equal 1. Neither standardisation is always applicable — the integral need not converge, and the supremum may fail to be finite, so that sometimes we have to select a convenient θ_o in the range of θ and require $L(\theta_o) = 1$. Even this may fail if, for some distributions to be considered, $L(\theta_o)$ should vanish. However, since only ratios of values of $L(\theta)$ are used, the standardisation is merely a matter of convenience. With the integral standardisation the likelihood function is identical with the density L appearing in the previous "Theorem".

Having written down the likelihood function, L, the method of maximum likelihood proceeds by differentiating $\ln L$ with respect to θ and equating the result to zero. Assuming that the root of the resulting equation is unique, it is called the maximum likelihood estimate of θ, and is denoted by $\hat{\theta}$. Thus, if

$$\frac{d}{d\theta} \ln L(\theta) = \lambda(\theta) ,$$

then

$$\lambda(\hat{\theta}) = 0 .$$

Next we differentiate $-\lambda(\theta)$ with respect to θ to obtain the information function

$$I(\theta) = -\lambda'(\theta) ;$$

the estimate is given in the form

$$\theta = \hat{\theta} \pm (1/\sqrt{I(\theta)})(N) .$$

(We have inserted the N to indicate that $1/\sqrt{I}$ is taken as the 'standard error' of the estimate θ, in the traditional sense.) The information function, $I(\theta)$, may be evaluated at $\hat{\theta}$, or its mean value evaluated under the hypothesis that $\theta = \hat{\theta}$. In any case, it is assumed that $I(\theta)$ does not depend strongly on θ, so that whichever alternative we adopt makes relatively little difference.

Now if $I(\theta)$ is approximately constant and equal to I, say, we have

$$\lambda(\theta) \approx \lambda(\hat{\theta}) - (\theta-\hat{\theta})I$$

from which

$$L(\theta) \approx K() \exp\{-\tfrac{1}{2}I(\theta-\hat{\theta})^2\}$$

since $\lambda(\hat{\theta}) = 0$. But this means that $L(\theta)$ is approximately a normal density centred at $\hat{\theta}$, with variance $s^2 = 1/I$. Referring to our "Theorem", we see that if $L(\theta)$ is approximately normal, the estimate t will be $\hat{\theta}$, and the scale factor s will be $1/\sqrt{I}$. Thus our method of conditioning will agree, approximately, with the method of maximum likelihood, the accuracy of the approximation depending upon the extent to which $I(\theta)$ is independent of θ . Before using the method of maximum likelihood in more general cases, when θ is not a location parameter, it is advisable to make a transformation of parameters in order to secure near constancy of $I(\alpha)$, where α is the transformed parameter. Of course, when θ is a location parameter, such a transformation will lead to a parameter which is not a location parameter, since any linear transformation, preserving location parameters, leaves $I(\theta)$ invariant.

To sum up, our method of conditioning will lead to approximately the same conclusions as the method of maximum likelihood, provided that the likelihood function is approximately normal, or, in other words, that the information function $I(\theta)$ is approximately independent of θ . If $L(\theta)$ is far from being normal the two approaches will give differing results.

Le Cam and Hajek, in a series of papers spread over many years, have obtained results which show that under very general conditions, as the sample size tends to infinity, all parameters tend to be satisfactorily approximated as location parameters, and all likelihood functions converge to normality.

Scale Parameters

We now proceed to consider the estimation of a scale parameter. We could do this directly, in terms of a scale parameter family with density

$$(1/\theta^n)\phi(x_1/\theta, x_2/\theta, \ldots, x_n/\theta)$$

but it will prove instructive if we introduce an unknown location parameter as well. Let the joint density of the observations $\underline{x}$ be

$$(1/\theta^n)\phi((x_1-\theta')/\theta, \ldots, (x_n-\theta')/\theta) \quad .$$

If we now define the pivotals $p_i = (x_i-\theta')/\theta$, they have joint density $\phi(\underline{p})$, and so satisfy the condition for being pivotals. We now transform $\underline{p}$ to $\underline{q}$ where

$$\begin{aligned} q_1 &= p_1 & \qquad & & p_1 &= q_1 \\ q_2 &= p_2-p_1 & & \text{i.e.} & p_2 &= q_1 + q_2 \\ q_r &= (p_r-p_1)/(p_2-p_1) & & & p_r &= q_1 + q_2 q_r \ , \end{aligned}$$

$$r = 3,4,\ldots,n \quad .$$

Again the transformation is 1-1 , but its Jacobian is now q_2^{n-1} . Thus the joint density of the q_i's is

$$q_2^{n-1}\phi(q_1, q_1+q_2, \ldots, q_1+q_2q_n) \quad .$$

Also it is easily seen that

$$q_r = (x_r-x_1)/(x_2-x_1) = y_r \ ,$$

say, so that q_r is known when the observations are known. It follows, by the conditionality principle, that the information in the data is equivalent to that in the observations x_1 and $d = x_2-x_1$, entering into the pivotals q_1,q_2 which have the joint conditional density

$$K()q_2^{n-1}\phi(q_1,q_1+q_2,\ldots,q_1+q_2y_n) \quad .$$

Now suppose we are told the value of θ' — say $\theta' = \theta_o'$. Putting $y_1 = x_1-\theta_o'$, we have

$$q_1 = y_1/\theta \quad , \quad q_2 = d/\theta$$

and by transforming q_1,q_2 to

$$q = q_2 \quad , \quad r = q_1/q_2 \qquad \text{(Jacobian} = q\text{)}$$

we find that we can now condition on the known value of $r = y_1/d$ and thereby obtain the conditional density for $q = d/\theta$,

$$K()q^n\phi(qr,q(r+1),\ldots,q(r+y_n)) \quad .$$

Now if we put

$$L(\theta) = \frac{\phi((x_1-\theta'_o)/\theta,\ldots,(x_n-\theta'_o)/\theta)/\theta^n}{\int \phi((x_1-\theta'_o)/\theta,\ldots,(x_n-\theta'_o)/\theta)(d\theta/\theta^n)}$$

the natural way of expressing our estimate is not in the form of an additive error, but rather in the form of a multiplicative error. In fact, $q = d/\theta$ has the density $L(1/q)$, and if we standardise this to L^* by the condition

$$\int \theta L^*(\theta)d\theta = 1$$

we can write our estimate as

$$\theta = t : s(L^*)$$

which should be read to mean that t/θ , multiplied by the scale factor s , has the standardised density L^* . The symbol ':' is used here as the multiplicative correlative of '$\pm$' . We shall have

$$t = \int zL(z)dz$$

and, since it is usual to scale densities on the positive real axis by their second moment about the origin,

$$s^2 = \int z^2L(z)dz \quad .$$

Regression

The essence of most of the preceding arguments can be found in Fisher's 1934 paper, though they appear there in connection with his fiducial argument which we have taken care to avoid. Similar arguments have also been produced by Fraser, in connection with his 'structural' theory of inference, though here again the theory is associated with a concept of parametric distributions (structural, instead of fiducial) which is not acceptable to everybody. In a recent paper Fraser and MacKay have extended their arguments to the general regression problem. Fisher and other early writers, however, seem to have failed to notice that Fisher's argument also extends to the problem of regression in a straightforward way, so long as we are prepared to consider the simultaneous estimation of all the unknown parameters entering the problem. It is when we wish to make statements about some parameters while regarding the other parameters as 'nuisance parameters' that further subtleties arise; these we shall discuss later.

We take the m rowed vector of observations $\underline{x}$ to be related to the n rowed vector of unknown (location) parameters $\underline{\theta}$ by

$$\underline{x} = g\,\underline{\theta} + \underline{e} \tag{8}$$

where g is a known $m \times n$ matrix of rank n and $\underline{e}$ is an m rowed vector of errors which is supposed to have density $(1/\sigma)^m \phi(\underline{e}/\sigma)$ where the density ϕ is known apart from the unknown scale parameter σ. Without loss of generality we may assume that the first n rows of g are linearly independent, so that

$$g = \begin{bmatrix} a \\ b \end{bmatrix}$$

with a square and non-singular. Then the first n equations, setting the errors to zero, can be solved to give the 'estimates'

$$\underline{\tilde{\theta}} = a^{-}\underline{x}$$

and if we write

$$\underline{x} = g\underline{\tilde{\theta}} + \underline{\tilde{e}} \tag{9}$$

the first n components of the residual vector $\underline{\tilde{e}}$ are zero, i.e.

$$\underline{\tilde{e}} = \begin{bmatrix} \underline{0} \\ \underline{f} \end{bmatrix} .$$

Subtracting (8) from (9) gives

$$\underline{0} = g\,(\tilde{\underline{\theta}} - \underline{\theta}) + \tilde{\underline{e}} - \underline{e}$$

so that
$$\underline{e} = g\,(\tilde{\underline{\theta}} - \underline{\theta}) + \tilde{\underline{e}}\ .$$

Since the transformation from $\underline{e}$ to the vector pivotals $(\tilde{\underline{\theta}} - \underline{\theta})/\sigma$ and $\tilde{\underline{e}}/\sigma$ is linear, its Jacobian, J, is constant and the joint probability function of $(\tilde{\underline{\theta}} - \underline{\theta})/\sigma$ and $\tilde{\underline{e}}/\sigma$ is $\phi(g\,(\tilde{\underline{\theta}} - \underline{\theta}) + \tilde{\underline{e}})J$. In order to obtain some known quantities having known densities on which to condition we introduce the pivotal z by

$$z^2 = \tilde{\underline{e}}'\tilde{\underline{e}}/\sigma^2 = r^2/\sigma^2 \quad \text{, say,}$$

and then set
$$f_i = z\Lambda_i \ , \quad i = 1,2,\ldots,m-n\ .$$

The vector $\underline{\Lambda}$ is then a known vector of direction cosines, and the Jacobian of the transformation from $\underline{p} = (\tilde{\underline{\theta}} - \underline{\theta})/\sigma$ and $\tilde{\underline{e}}$ or $\underline{f}$ to $\underline{p}$, z , and $\underline{\Lambda}$ is $z^{m-n}F(\underline{\Lambda})$, where $F(\underline{\Lambda})$ is a product of sines and cosines which we need not evaluate. The joint density of $\underline{p}$ and z , conditioned on the observed values of $\underline{\Lambda}$ will then be

$$\xi(\underline{p},z) = K()z^{m-n}\phi(g\,\underline{p} + \begin{bmatrix}\underline{0}\\ z\underline{\Lambda}\end{bmatrix})\ .$$

If, for example,

$$(1/\sigma)^m\phi(\underline{e}/\sigma) = \{1/(\sigma\sqrt{2\pi})\}^m|v|^{-\frac{1}{2}m}\exp\{-\tfrac{1}{2}\underline{e}'v^-\underline{e}/\sigma^2\}$$

we find that

$$\xi(\underline{p},z) = K()z^{m-n}/\sigma^m\exp[-\tfrac{1}{2}\{\underline{p}'g'v^-g\underline{p} + z^2(\underline{\Lambda}'h_{22}\underline{\Lambda}) + 2z(\underline{0}',\underline{\Lambda}')v^-g\underline{p}\}]$$

where h_{22} is the appropriate-sized lower diagonal block of v^-, giving as marginal density for $\underline{p}$ the n-variate normal density with variance matrix $g'v^{-}g$ and mean vector

$$(g'v^{-}g)^{-}g'v^{-}\begin{bmatrix} 0 \\ z\underline{\Lambda} \end{bmatrix} = \underline{\hat{\theta}} - \underline{\theta}$$

where $\underline{\hat{\theta}}$ is the 'generalised least squares estimate' of θ

$$\underline{\hat{\theta}} = (g'v^{-}g)^{-}g'v^{-}\underline{x} \; .$$

If $(g'v^{-}g)^{-} = ss'$ and $\underline{\hat{e}} = \underline{x} - g\underline{\hat{\theta}}$ then we can derive the estimates

$$\underline{\theta} = \underline{\hat{\theta}} \pm \sigma\, s(N) \qquad (10)$$

and

$$\sigma^2 = \frac{\underline{\hat{e}}'v^{-}\underline{\hat{e}}}{m-n} : (\chi^2_{[m-n]}/(m-n)) \; . \qquad (11)$$

The second estimate, (11) , is interpreted to mean that the density of the pivotal $(\underline{\hat{e}}'v^{-}\underline{\hat{e}})/(m-n)\sigma^2$ is that of a chi-squared distribution with m-n degrees of freedom, normalised to have unit mean. Correspondingly the first estimate, (10) is to be understood to mean that the vector pivotal $(s^{-}/\sigma)(\underline{\hat{\theta}}-\underline{\theta})$ has a standard n-variate normal distribution. This vector pivotal and the pivotal for σ^2 are distributed independently, so that from the two separate estimates we can derive the joint density of the pivotals, if required. We could treat the estimates as altogether independent of each other if it were not for the fact that σ, which is estimated in the second expression, enters into the first estimate. A further problem arises if we wish to estimate one component of $\underline{\theta}$, for example the first, θ_1 . While the estimate derived from the marginal distribution of $(\hat{\theta}_1-\theta_1)$, viz.

$$\theta_1 = \hat{\theta}_1 \pm \sigma s_{11}(N) \qquad (12)$$

expresses all the *direct* information we have concerning θ_1, it may well be that the off-diagonal elements s_{1i}, $i \neq 1$ of s are far from negligible, and that any information which later came to hand about the other components of $\underline{\theta}$ could be highly informative concerning θ_1 — it could almost determine θ_1 uniquely. Thus if we use the estimate (12), it should be on the understanding that *nothing is known, apart from the present experimental data, concerning the possible values of the other components of* $\underline{\theta}$.

It may be remarked that one advantage of orthogonal designs of experiments is that the resulting estimates of the vector of unknowns then have s in diagonal form, so that the particular difficulty just noticed does not arise. It may also be the case that instead of estimating the vector $\underline{\theta}$, we may be prepared to estimate a transform $t\underline{\theta} = \underline{\alpha}$, say, where t is a non-singular matrix. If t can be chosen so that ts is diagonal, again the difficulty can be avoided.

The problem of the unknown σ entering (10) is in some ways similar to the problem of possible relatedness of the estimates of components of $\underline{\theta}$. Obviously, if σ were known exactly, we would use its known value in (10), and disregard (11). Similarly, if other information came to hand concerning σ, beyond that given in (11), it should be used in interpreting (10). We shall show later how this can be done in some cases. Meanwhile there remains the question of what should be done if there is no information about σ apart from that provided by the experiment.

It was, of course, the special case when $v=I$, $g=\underline{1}$ that 'Student' considered in his famous 1908 paper on 'The Probable Error of a Mean'. His solution amounted, in our terms, to the estimate

$$\theta = \hat{\theta} \pm \hat{\sigma}\, s_{11}(t_{m-1})$$

where

$$\hat{\sigma}^2 = (\underline{\hat{e}}'\underline{\hat{e}})/(m-1) \quad , \quad s_{11} = 1/\sqrt{m}$$

and (t_{m-1}) refers to Student's t distribution on $m-1$ degrees of freedom.

We can obtain the result corresponding to Student's, from our estimates (10) and (11), by further transforming the pivotals so as to 'separate the parameters', i.e. to obtain a set of pivotals each of which contains just one parameter. Writing

$$z^{*2} = (\underline{\hat{e}}'v^{-}\underline{\hat{e}})/(m-n)\sigma^2 \quad ,$$

we can divide the pivotal $(\underline{\hat{\theta}}-\underline{\theta})/\sigma$ by z^* to obtain

$$\underline{u} = (\underline{\hat{\theta}}-\underline{\theta})/w \quad ,$$

where

$$w^2 = (\underline{\hat{e}}'v^{-}\underline{\hat{e}})/(m-n)$$

and the joint density of $\underline{u}$ and z^* will be given by

$$K()(z^*)^m \exp[-\tfrac{1}{2}w^2\{\underline{u}'g'v^{-}g\underline{u}+(m-n)\}] \quad . \qquad (13)$$

If we use the marginal distributions of $\underline{u}$ and of z^* to form our estimates we obtain

$$\underline{\theta} = \underline{\hat{\theta}} \pm w\ s\ (t_{m-n}) \qquad (14)$$

and

$$\sigma = w : (\chi^2_{[m-n]}/(m-n)) \qquad (15)$$

but we must remember that in using the marginal density, for example, of $\underline{u}$, we are assuming the validity of the joint distribution, and hence that of z^*.

That is to say, we continue to suppose that nothing is known concerning the value of σ beyond what we learn from the data.

If we intend to use the estimate (14), the question arises as to how we are to express the fact that this is 'conditional on the estimate (15)', as it were. Fisher's answer to this question was that we should interpret (14) in terms of a reference set of repeated experiments in which the estimate w is fixed, the same in each experiment, but the pivotal z^* continues to have its appropriate distribution, so that σ is understood to vary from one experiment to another. The distribution of σ so generated was the fiducial distribution of σ, according to Fisher, but its application in the present context by no means commits us to general acceptance of the fiducial argument. In the present context, the distribution of σ may be regarded as a hypothetical one, introduced in order to give expression, in the interpretation of (14), to the fact that (14) depends for its validity on the absence of knowledge of σ beyond that provided by (15). Stated in another way, this use of the fiducial distribution in connection with nuisance parameters (as σ would be called, if the parameters of interest were just $\underline{\theta}$) in no way commits us to acceptance of the full fiducial argument, and in particular does not expose us to the argument concerning betting strategies which has been sketched previously. We are simply accepting a convention whereby we interpret the information about parameters of interest in terms of a reference set in which the nuisance parameters are imagined to vary so as to maintain the pivotal distributions, while the functions of the observations entering the pivotals for the nuisance parameters -- the estimates of the nuisance parameters -- remain fixed.

There are at least two somewhat compelling technical arguments in favour of such a convention as against, for example, the 'Neyman' convention, which specifies that the reference set must necessarily consist of repetitions of the sampling procedure from the same population. These are in addition to the point that in most scientific contexts the sampling procedure is not at all easy to define -- it often is essentially conditioned by the availability of funds and of opportunities for observation.

The first technical point arises from the fact that if we regard the reference set in Student's t situation, for example, as consisting of repeated samples of a fixed size from one fixed normal population, then the interpretation of the confidence interval statements derived from the t distribution of the pivotal

$$t = (\bar{x} - \theta)\sqrt{n}/s$$

becomes difficult. The usual interpretation is that we can be sure that, in the long run of statements such as

$$\bar{x} - t_{0.9}s/\sqrt{n} < \theta < \bar{x} + t_{0.9}s/\sqrt{n}$$

where $t_{0.9}$ is the value of t which cuts off 90% of the distribution, 90% of these statements will be true. This is correct. But it is possible to select from this long run of statements a subset for which we can be sure that at least 93% will be true, for instance, while among the remainder of the original sequence, less than 90% will be true. We do this by selecting the first subsequence by the condition that

$$(\bar{x} - a)\sqrt{n}/s < C$$

where a is an arbitrary constant, and C is a constant suitably chosen.

This will select from among the original sequence one in which the observed values of s are rather larger than usual, and since the width of the confidence interval is proportional to s, it is not unreasonable to expect that the convergence probability should be larger than usual; computation shows that this is indeed the case. The presence of the arbitrary constant a is particularly awkward since it means that for any particular statement which we have to hand, we can choose a so that this statement belongs to the selected set. Having once chosen a, of course, it must be held constant for all the remaining statements. Similarly, it will be possible to choose a so that a given statement will fail to belong to the selected subset. Thus, it is hard to see, with this interpretation, how it could be reasonable to attach the 'confidence' 90% to any particular statement -- unless we are prepared to introduce a convention which forbids the selection of subsets of the kind discussed. But as soon as we do this, we depart from the pristine simplicity of the Neyman reference set. While we would not suggest that there is any mathematical inevitability about the convention of fixing the estimate of s and allowing σ to vary according to the pivotal distribution, such a convention seems more reasonable when regarded in this light. The Fisherian reference set can be interpreted as meaning that the value we have observed for s is typical -- neither unusually large nor unusually small.

The second technical point arises when we proceed to consider two normal populations, with one sample from each. We assume that the four parameters are allowed to vary freely, and that the parameter of interest happens to be δ, the difference of the two population means; $\bar{\mu}$, the mean of these means, and the population standard deviations σ_1, σ_2 are nuisance parameters. Under these conditions it turns out that although all the information about all the parameters can be summed up in the sufficient statistics $\bar{x},\bar{y},s_1,s_2$, the sample means and sample standard deviations, no function of these four quantities can be found to provide an estimate of δ; more precisely, no pivotal function $f(x,y,s_1,s_2,\delta)$ exists whose distribution, in repeated sampling from the same population, does not involve any parameter. On the other hand, if we adopt the Fisherian convention for the reference set, and allow

$$\tan\theta = (s_1^2/s_2^2)(n_2/n_1) \qquad \text{(where } n_1,n_2 \text{ are the sample sizes)}$$

to be regarded as fixed, while the pivotal

$$(s_1^2/s_2^2)(\sigma_2^2/\sigma_1^2)$$

has its appropriate distribution generated by variation of the ratio σ_1^2/σ_2^2, then it turns out that we can provide the natural estimate

$$\delta = d \pm s(BFS(\theta))$$

where $d = \bar{x} - \bar{y}$, $s = \sqrt{\frac{s_1^2}{n_1} + \frac{s_2^2}{n_2}}$ and $BFS(\theta)$ denotes the Behrens-Fisher-Sukhatme distribution with the parameter θ. This estimate will have the usual interpretation that $(d-\delta)/s$ has, in the reference set in which θ is fixed, the $BFS(\theta)$ distribution. The result presented here is due essentially to Linnik. It can be generalised, and to some extent simplified, in the following way.

We suppose we have one sample from each of two populations (not necessarily normal) with location and scale parameters $\mu_1,\mu_2,\sigma_1,\sigma_2$. By the methods already indicated we can derive fully efficient estimates for the four parameters:

$$\mu_1 = m_1 \pm s_1(D_1^*) \quad , \qquad \mu_2 = m_2 \pm s_2(D_2^*) \quad ,$$

$$\sigma_1 = s_1 : k_1(E_1^*) \quad , \qquad \sigma_2 = s_2 : k_2(E_2^*) \quad .$$

We wish to derive an estimate for $\delta = \mu_1 - \mu_2$. By regarding the two 'standard errors' s_1, s_2 as fixed, from the estimates for μ_1 and μ_2 we can derive the estimate

$$\delta = (m_1 - m_2) \pm s\ (D^*)$$

where the 'standard error' $s = \sqrt{s_1^2 + s_2^2}$ and the distribution D^* is obtained as the convolution of D_1^* and D_2^* . It turns out that the resulting distribution depends only on the ratio s_1/s_2 ; therefore, all we need specify about the conventional reference set involved in the distribution D^* is that it is one in which this ratio is fixed and the ratio of the population standard deviations is assumed to vary in such a way as to preserve the distribution of the pivotal

$$s_1\sigma_2/s_2\sigma_1 \quad .$$

Pivotal Inference in General

Let us now sum up the preceding arguments. We have been dealing with situations in which we have observations and parameters linked by means of *pivotals*, that is to say, functions of observations and of parameters whose distribution is fixed, independent of the parameters. Mathematically, if we call the set of possible observations the sample space S , and the set of possible parameter values the parameter space Ω , a pivotal is a function from $S \times \Omega$ to some Euclidean space E , and the pivotal possesses a determinate distribution over E . It is helpful to define a *Bayesian pivotal* as one which is constant on S — that is, it is a function only of parameters, not of observations. Specifying the distribution of a Bayesian pivotal is equivalent to assuming a 'prior' distribution, in the sense of Bayes' Theorem, for the parameter(s) involved. Dual to the concept of a Bayesian pivotal is the concept of an *ancillary pivotal*, one which is constant on the parameter space Ω . It constitutes a function of the observations whose distribution is known. A pivotal which is neither Bayesian nor ancillary can be called *proper*.

We specify the model for our experiment in terms of pivotals. Whenever possible, we transform these into ancillary pivotals and then condition the distribution of the remaining pivotals on the observed values of the ancillaries. If all the remaining pivotals are Bayesian, these will generate a joint posterior distribution for the parameters which solves the general *joint* estimation problem. If some of the parameters are nuisance parameters, the marginal distribution of the parameters of interest may be used as the solution of the general *separate* estimation problem, subject to the proviso that no information is available or may come to hand concerning the values of the nuisance parameters other than that derived from the given experiment.

In general, complete separation of pivotals into ancillaries and Bayesians will not be possible, and we shall be left with some proper pivotals. If it is possible to find a single proper pivotal containing a parameter of interest, distributed independently of all the remaining pivotals, then the information about this parameter can be regarded as contained in this proper pivotal with its distribution. If the standardised pivotal is of the form $(t-\theta)/s$, then the information is expressible as the estimate

$$\theta = t \pm s(D^*)$$

where D^* is the (standardised) distribution of the pivotal $(t-\theta)/s$. If the standardised pivotal is of the form t/θ , the information is expressible as the estimate

$$\theta = t : s(D^*)$$

We discuss below what happens if the pivotal takes neither of these forms.

The general case will be one in which complete, independent separation of the parameters of interest is not possible. In this case we either have to resort to approximations, or we have to adopt conventional reference sets in which the nuisance parameters are understood to follow distributions such that the distributions of the pivotals containing these nuisance parameters are preserved while their estimates from the observations are fixed.

If we wish to deal with altogether general situations, the preceding discussion underlines the truth of Fisher's often repeated point that the nature of the inferences we can draw from a set of data depends crucially on the mathematical form of the model specified. On the other hand, our treatment of the general linear regression problem shows that in many practical cases the situation is not as bad as mathematical generality might suggest, especially if we are prepared to accept reasonable approximations.

We may remark here, that the demand for *separate*, as opposed to *joint* estimates is often made unthinkingly, and estimates which are of necessity joint are sometimes unthinkingly interpreted as if they were separate. A case in point is provided by the sample from a normal distribution with unknown mean μ and unknown standard deviation σ, with sample size n, mean $\bar{x}$, and sample standard deviation s. The pivotals

$$t = (\bar{x} - \mu)\sqrt{n}/s$$

$$z = s/\sigma$$

are not statistically independent, so that confidence intervals for μ derived from the marginal distribution of t involve appeal to a conventional reference set in which σ is understood to vary in a particular way. On the other hand the pivotals

$$p = (\bar{x} - \mu)\sqrt{n}/\sigma$$

and

$$z = s/\sigma$$

are statistically independent. The V-shaped confidence set produced in the (μ,σ) plane by taking upper and lower limits for p highlights very clearly that what we can say about μ depends on what we know about σ. And yet we can find, at the same time, an *independent* confidence set for σ by using the pivotal z; this confidence set takes the form of a strip in the (μ,σ) plane parallel to the μ axis. Finally, since p and z are independent, if the V-shaped set has confidence coefficient C and the strip has confidence coefficient C', the wedge-shaped area in which they intersect will be a confidence region with confidence coefficient CC'. Such a region will often be more informative than the vertical strip to which the use of t leads.

An extreme case of misinterpretation arises in the preceding example if the sample size is 1 . In this case only the pivotal p can be defined, and it is sometimes asserted that the 'maximum likelihood' estimates for μ and σ are the 'absurd' values, x and 0 , respectively. Regarded as separate estimates these values are indeed absurd; regarded as joint estimates, however, they have a sensible interpretation.

Robust Pivotals — Robust Estimation

In our general description of the pivotal inferential process it has been taken for granted that the description of a statistical model in terms of pivotals is unique. That this is not so can be seen from the multivariate normal case, where to say that the k-dimensional vector of observations $\underline{x}$ has a spherical normal distribution around the vector mean $\underline{\mu}$ we can specify either that the vector pivotal

$$\underline{p} = \underline{x} - \underline{\mu}$$

has the standard multivariate normal distribution, or that the scalar pivotal

$$d^2 = (\underline{x}-\underline{\mu})'(\underline{x}-\underline{\mu})$$

has a χ^2 distribution on k degrees of freedom independently of the vector

$$\underline{a} = (\underline{x}-\underline{\mu})/d$$

which is assumed to have a uniform distribution over the unit k-sphere. Now in practice, with real observations, we shall almost never have precise knowledge of the pivotal distributions. When we say 'normal' we almost always mean 'approximately normal', and quite often our real information is very vague indeed. It is easy to see that the equivalence of our two modes of specifying the k-variate normal distribution depends critically on the form of the population assumed. Quite small departures from normality of the individual observations, for example, could produce marked departures from uniformity in the distribution of $\underline{a}$, while departures from the χ^2 distribution for d^2 , though retaining uniformity for $\underline{a}$, would produce dependence among the components of $\underline{x}$.

Often, therefore, the way to proceed in model specification is to state what quantities are to be considered as pivotal, and then to carry through the analysis for a range of distributional forms. It may or may not turn out that the inference finally drawn depends critically on the form of distribution originally assumed. As we have previously remarked, for example with samples from a Cauchy distribution, we often find that the final inferences about location (and, less often, scale) are very similar to those which we would draw on assuming a normal distribution. If, for a given sample, this were shown to be the case, then we could say that *for the sample lying to hand* the inferences were robust with respect to such changes in distributional form. However, if the inferences did change materially according to the distributional form assumed, then it would be well to know this, and to make inferences for the given sample conditional on the assumed form or forms.

This approach to robust estimation differs from that of Andrews *et al.* who attempt to find functions of the data whose relationship to the parameters is in some sense approximately independent of the assumed distributional form. Such an approach fails to acknowledge that the distributional form sometimes will, and sometimes will not matter, and that it is possible to discriminate between the two types of situation by examining characteristics of the observed sample.

Categorical Data

In the preceding discussion the statistical model is best thought of as being specified first of all by a set of pivotals relating the observations to the parameters, and then by the distribution of these pivotals; moreover, specification of the distributional form may require wise uncertainty. We now pass to the consideration of a contrasting set of situations where theory specifies a particular form for the probability distribution. Such cases commonly occur in genetics where the frequencies of occurrence of phenotypes are specified in terms of recombination fractions and similar parameters. Likewise, in high energy physics, for example, a measured angle between two particle tracks is such that its cosine, x , has the probability density

$$\phi(x) = (1 + \theta x)/2$$

with θ restricted to lie between -1 and $+1$.

Cases of this kind can be brought under the rubric of pivotal inference only approximately. In large samples, where n_i is the observed number of items in the ith category, and $p_i(\underline{\theta})$ gives the probability for this category as a function of the unknown parameters $\underline{\theta}$, then the quantity

$$(n_i - Np_i) / \sqrt{Np_i(1-p_i)}$$

can be taken to have approximately a standard normal distribution, and the set of all such quantities will be approximately jointly normal with a variance matrix which depends on N , the total number of items observed. But for moderate sample sizes another approach is desirable, and becomes essential for small samples.

Such an approach is provided by the *score function*. This is defined as the gradient of the logarithm of the likelihood function. If $\phi(x,\underline{\theta})$ expresses the probability function of the observations x in terms of the unknown parameters $\underline{\theta}$, the score function is the vector-valued function $\underline{S}$ whose i^{th} component is

$$S_i(\underline{\theta};x) = \frac{\partial}{\partial\theta_i} \ln \phi(x,\underline{\theta}) ,$$

and its observed value in a given experiment is this function of $\underline{\theta}$ evaluated for the observed values of x. It is, therefore, a function-valued function of the observations. Associated with the score function is the *information matrix*

$$I(\underline{\theta};x) = -\{\partial S_i/\partial\theta_j\} .$$

It should be noted that the information function can be determined if the score is known for all values of $\underline{\theta}$. Thus the information function does not give us information beyond that contained in the score function. Similarly, the score function determines the likelihood function, and in this case the converse is true -- the likelihood function determines the score function.

When Fisher wrote *Statistical Methods for Research Workers* interactive computing was non-existent or, for the later editions, in its infancy. Thus in that book a lot of emphasis is placed on approximations to the score function, mainly based on the fact that it is usually approximately linear in the unknown parameters. However modern pocket calculators can even be programmed to compute the score function for any proposed value of $\underline{\theta}$, so that the original emphasis on linear approximations is now unnecessary and complete tabulation of the score function can be regarded as practicable. An indication of Fisher's moving in this direction can be found in section 57.4 which was added in the last edition of his book.

The fundamental property of the score function which gives rise to its importance in inference is that it is equivalent to the minimal sufficient statistic. Recall that a function $T(x)$ of the observations is said to be sufficient for a parameter θ if the conditional probability function of the observations, given the value of T, does not involve the parameter θ. When we know the value of a sufficient statistic, no further information gleaned from the observations is relevant to θ. A minimal sufficient statistic M is a sufficient statistic with the property that any sufficient statistic T must be determined by M; in other words, there must be a single valued function of M such that $T = f(M)$. If M and M' are both minimal sufficient, we must have $M = f(M')$ and $M' = f'(M)$ — that is, there must be a one-to-one correspondence between the values of M and those of M', so that these two statistics are in that sense equivalent to each other; hence we speak of *the* minimal sufficient statistic.

Not only is $\underline{S}$ a fully efficient summary of the data, it is also a reasonably convenient one in many cases. The value of $\underline{S}$ at the true parameter point, say $\underline{\theta}_o$, is a numerically-valued statistic whose mean value is always

$$\langle \underline{S}(\underline{\theta}_o;x) \rangle_{\underline{\theta}_o} = \underline{0} .$$

Thus if we solve the equation

$$\underline{S}(\underline{\theta};x) = \underline{0}$$

for $\underline{\theta}$, we shall, if the solution is unique and represents a possible value for $\underline{\theta}$, obtain a vector $\hat{\underline{\theta}}$ which can be considered an estimate of $\underline{\theta}$ in the sense that it is obtained by equating to zero a quantity whose mean value is zero. In this case $\hat{\underline{\theta}}$ is called the maximum likelihood estimate. It is also the case that the variance matrix of $\underline{S}(\underline{\theta}_o;x)$ is

$$v(\underline{S}) = \langle I(\underline{\theta}_o;x)\rangle_{\underline{\theta}_o}$$

so that if we factorise v as ss' , the quasi-pivotal

$$\underline{\lambda} = s^{-}\underline{S}$$

will have zero mean and its variance matrix will be I . We call it quasi-pivotal because, like a pivotal, its mean and its variance are independent of the unknown parameters, though its higher moments may not be. If all its moments were independent of the unknown parameters it would be fully a pivotal.

When θ is a scalar it can be shown that if g is any other quasi-pivotal, having mean zero and unit variance in common with λ , then

$$\text{cov}(g,\lambda) = -\frac{1}{s}\langle dg/d\theta\rangle$$

and since this covariance cannot in modulus exceed 1 , while if g coincides with λ it will equal 1, it follows that

$$|\langle dg/d\theta\rangle| \leq |\langle d\lambda/d\theta\rangle| \quad .$$

If, on the basis of Chebychev's inequality, we attach a probability of at least 8/9 to the proposition $|\lambda|\leq 3$ or to the proposition $|g|\leq 3$, then since λ and g are approximately linear in θ , the confidence limits thus derived for θ will be closer using λ than using g . In this sense λ is an optimal quasi-pivotal. A more complex corresponding result holds in the case when θ is a vector.

Clearly, much depends on the approximate linearity of $\underline{S}$ — that is, on the approximate constancy of the matrix $I(\underline{\theta};x)$ over the range of values of $\underline{\theta}$ reasonably consistent with the data. A feature of the location parameter case discussed earlier which accounts for the particularly effective way in which that case can be handled is that it is possible to pick out a subset of samples over which the information takes exactly the same form for the whole range of values of the parameter. In these cases the information matrix is not constant as a function of $\underline{\theta}$, and so $\underline{S}$ is not linear.

To conclude, we return briefly to the general question of reference sets; the intuitively appealing idea behind all of them seems to be that we should relate the sample we have to other samples which are "equally informative". One way of approaching this is to require that the reference set should consist of samples of the same size, even though a careful study of the experimental procedure used would lead to the conclusion that, in repetitions of the procedure, continued use of exactly the same sample size would be most unlikely. If we refer to the example quoted in relation to two alternative conditionings (the accident rates for a car of given cubic capacity and given make), and to the mathematics contrast there made between the normal and the rectangular distribution, it can be seen that no single, satisfactory measure of 'information' is likely to be devised.

REFERENCES

Andrews, D.F., Bickel, P.J., Hample, F., Huber, P.J., Rogers, W.H. and Tukey, J.W. (1972) *Robust Estimates of Location.* Princeton: Princeton University Press.

Barnard, G.A. (1976) Conditional Inference is not Inefficient. *Scandinavian Journal of Statistics, 3,* 132-134.

Bliss, C.I. (1967) *Statistics in Biology, Vol. 1.* New York: McGraw-Hill Book Company, Inc.

Fisher, R.A. (1934) Two New Properties of Mathematical Likelihood. *Proceedings of the Royal Society, A144,* 285-306.

Fisher, R.A. (1925-70) *Statistical Methods for Research Workers.* (14th edition) New York and London: Hafner Press.

Fraser, D.A.S. (1976) Necessary Analysis and Adaptive Inference. *Journal of the American Statistical Association, 71,* 99-113.

Fraser, D.A.S., and MacKay, R.J. (1974) Parameter Factorisation and Inference Based on Significance, Likelihood and Objective Posterior. *Annals of Statistics, 3,* 559-572.

CONDITIONAL INFERENCE IS NOT INEFFICIENT

G.A. BARNARD

Reprinted with the permission of the
Scandinavian Journal of Statistics

ABSTRACT. A paper of Welch (1939) is often referred to as demonstrating the inefficiency of conditional confidence intervals. It is the object of this note to show that such a notion arises from a misunderstanding.

Following Welch, for simplicity we restrict consideration to an unknown parameter of location θ to be estimated from a sample $x_1, x_2, \ldots, x_n$ of n independent and identically distributed observations with common density $\phi(x_i - \theta)$. The conditional argument takes any convenient location parameter of the sample, such as the sample mean $\bar{x}$, along with $n-1$ further statistics, such as $c' = (c_1, c_2, \ldots, c_{n-1})$ where

$$c_i = x_i - \bar{x} \qquad i = 1,2,\ldots,n$$

and transforms the data to

$$\begin{aligned} \bar{x} &= \textstyle\sum x_i/n \\ c_1 &= x_1 - \bar{x} \\ &\ldots \\ c_{n-1} &= x_{n-1} - \bar{x}\,. \end{aligned}$$

The transformation is $1-1$, and so information preserving, and its Jacobian J is constant. Thus the joint density

$$\prod_i \phi(x_i - \theta)$$

transforms to

$$J \prod_i \phi(\bar{x} - \theta + c_i) .$$

The conditioning argument then implies that, since the c_i are known, the information concerning θ can be concentrated in $\bar{x}$ with its conditional density

$$\eta(\bar{x} - \theta | c) = \prod_i \phi(\bar{x} - \theta + c_i)/\xi(c) \qquad (1)$$

where

$$\xi(c) = \int \prod_i \phi(p + c_i)dp \qquad (2)$$

is the marginal density of c.

From this conditional density we can obtain conditional confidence intervals for θ based on the pivotal $p = \bar{x} - \theta$. Specifically, the set

$$\{\theta:(\bar{x} - \theta) \in I\} \qquad (3)$$

is a $1 - \alpha$ confidence set for θ if

$$\int_I \eta(p|c)dp = 1 - \alpha . \qquad (4)$$

We minimise the measure $\ell(I)$ of such a set I by taking it to be of the form

$$I = \{p:\eta(p|c) \geq k(\alpha,c)\} \qquad (5)$$

where $k(\alpha,c)$ is chosen to satisfy (4). (The trivial complications arising when $\eta(p|c)$ is constant and positive on a set of positive measure are ignored.) Then for future reference we note that

$$\frac{d\ell}{d\alpha} = -k(\alpha,c). \qquad (6)$$

Now suppose we wish to minimise the mean length of our confidence intervals:

$$\int \ell(\alpha,c)\xi(c)dc \tag{7}$$

subject to the overall confidence coefficient being $1 - \bar{\alpha}$. For this last requirement we must have

$$\int \alpha(c)\xi(c)dc = 1 - \bar{\alpha} \tag{8}$$

and Welch correctly points out that to minimise (7) subject to (8) we should not in general take

$$\alpha(c) = \bar{\alpha}.$$

In fact (7) and (8) define a variational problem for which, introducing the undetermined multiplier λ, we require the variation

$$\delta\int \{\ell(\alpha(c),c) - \lambda\alpha(c)\}\xi(c)dc$$

to vanish. The condition for this is

$$\frac{\partial \ell}{\partial \alpha} = \lambda$$

from which, using (6), we have

$$k(\alpha(c),c) = -\lambda \,. \tag{9}$$

This implies that, to obtain confidence intervals with overall coefficient $1 - \bar{\alpha}$ and, subject to this, minimum average length, we should select a set S in the n-dimensional space (p,c) such that in S,

$$\eta(p|c) = \frac{\prod\limits_i \phi(p + c_i)}{\xi(c)} \geq -\lambda \tag{10}$$

where λ is chosen so that

$$\int\left\{\int_S \eta(p|c)dp\right\}\xi(c)dc = 1 - \bar{\alpha} \tag{11}$$

i.e. so that

$$\int_S \cdots \int \prod_i \phi(p + c_i)dpdc_1 \ldots dc_{n-1} = 1 - \bar{\alpha} \; . \tag{12}$$

The procedure for deriving conditional confidence intervals from this set is illustrated by the following example, which again follows Welch. We suppose two independent observations x_1, x_2 are available from the rectangular distribution with density $U(x_i - \theta)$, where

$$U(z) = 1, \quad |z| \le \frac{1}{2}$$

$$= 0, \quad \text{otherwise.}$$

In this case we have

$$p = ((x_1 + x_2)/2) - \theta$$

$$c = (x_1 - x_2)/2$$

and the joint density of (p,c) is

$$2U(p + c)U(p - c) \tag{13}$$

which is 2 in the square

$$|p + c| \le \frac{1}{2} \text{ and } |p - c| \le \frac{1}{2} \tag{14}$$

and zero elsewhere. We have in this case

$$\xi(c) = 4\{\tfrac{1}{2} - |c|\}, \quad |c| \le \frac{1}{2} \tag{15}$$

$$= 0, \quad \text{elsewhere}$$

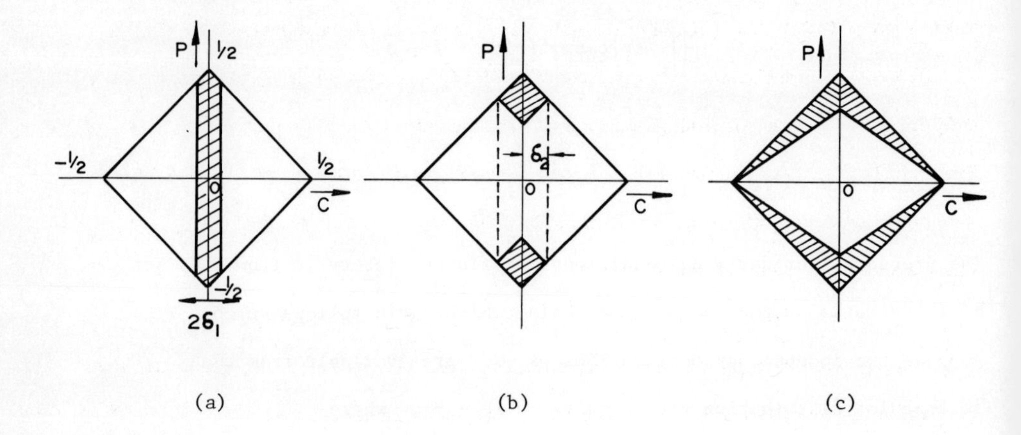

Figure 1

and so the region S in this case is obtained by deleting from the square the strip

$$|c| < \delta_1$$

where $2\left(\frac{1}{\sqrt{2}} - \delta_1\sqrt{2}\right)^2 = 1 - \bar{\alpha}$, i.e. where $\delta_1 = \frac{1 - \sqrt{1 - \bar{\alpha}}}{2}$.

This gives confidence intervals of zero length where $|c| < \delta_1$, and intervals of length $1 - 2|c|$ when $|c| > \delta_1$. The region S is shaded in Figure 1a.

Welch's proposal was to take the region S', shaded in Figure 1b, which he showed was preferable to the shaded area in Figure 1c. From Figure 1b we must have

$$8\delta_2{}^2 = \bar{\alpha}$$

and the confidence intervals in this case are of length $1 - 2|c|$ when $|c| > \delta_2$, but when $|c| \leqslant \delta_2$ they are of length $\{1 - 4\delta_2 + 2|c|\}$.

Taking $\bar{\alpha} = 0.1$, the mean length for confidence intervals based on S is 0.56921 , while the mean length for Welch's confidence intervals is 0.57412 . The mean length of the intervals using $\alpha(c) = \bar{\alpha}$ is 0.6. Thus while Welch's intervals are, on average, shorter than those in which the confidence coefficient is fixed, irrespective of the conditioning set, they are longer than those based on S.

The logic of the situation would seem to be best put as follows: Given the sample data, converted into the pivotal p, and the conditioning ancillaries c, we base confidence intervals on the conditional distribution $\eta(p|c)$. The confidence coefficient $1 - \alpha(c)$ we choose can depend on c. If we choose the dependence on c to be such, averaged over c, as to equal $\bar{\alpha}$, then we can talk of an averaged confidence coefficient $1 - \bar{\alpha}$. If we then pose the problem of minimising length, or the probability of covering false values, or any other property, subject to a given $\bar{\alpha}$, we can in principle choose our function $\alpha(c)$ accordingly. But *having obtained* an observed value of c, the relevant confidence coefficient will be $\alpha(c)$, not α. Otherwise we find ourselves in the absurd position of sometimes asserting, with 90% confidence, that something almost certainly false is true, and at other times asserting, with only 90% confidence, that something is true which we know for certain is so. In any case there seems no justification for the bald assertion that conditional confidence intervals are inefficient. To obtain efficient confidence intervals overall we should adjust our confidence coefficients to the precision of our measurements - as Cox pointed out in 1958.

Reference Welch, B.L. (1939). On confidence limits and sufficiency, with special reference to parameters of location. *Ann. Math. Stat.* *10*, 58-69.

Bioassay

Professor D.J. Finney

Department of Statistics
University of Edinburgh
Edinburgh, Scotland

BIOASSAY

Lecture I: GENERAL STRUCTURES AND THEORY

1. Purpose and Statistical Implications

The types of bioassay with which I am concerned are experimental procedures for estimating the biological potency of a material such as a therapeutic drug or pesticide. They are used *either* because chemical or physical analysis is inadequate to assess a specific biological potency (as is typically the case with materials that are themselves biological in origin) *or* because, as compared with chemical analysis and in relation to practical requirements of potency estimation, the assay has advantages of speed, cost, or other important features.

Bioassay techniques have been used for many years and for a wide range of purposes. The *Book of Genesis* records how Noah used birds to assess the extent to which flood waters had declined; one may suspect that this story of a necessarily crude and inexact measurement had some basis in fact. Certainly from early in this century pharmacologists, toxicologists, and others have employed tests on animal tissue for making numerical comparisons between different agents possessing similar biological activities. This work began, and doubtless could continue, without the aid of statisticians! Why then should we interest ourselves in the subject? I suggest three reasons:

(1) Because they depend upon types and magnitudes of biological reactions that vary from subject to subject, assays tend to be inherently imprecise. Statisticians can assist in the numerical assessment of precision; they can devise experimental designs for the improvement of precision, and thus contribute to economy in the use of resources.

(2) Particular types of assay produce problems of statistical theory

and method, full or approximate solution of which may be essential to good assay practice.

(3) The planning, analysis and interpretation of bioassay experiments provide excellent illustrations of an integrated approach to biometric practice; typically there is need to consider problems of estimation theory in relation to experimental design, and the distinct roles of hypothesis testing and estimation can be clearly seen.

This first lecture is largely based upon my own book (Finney, 1964).

2. The General Model

At one time, biological potency was occasionally measured in terms of quantity required to produce an effect of specified magnitude in an animal or other biological system. This parallels early practices with other kinds of measurement; for example, such measurements of length as the foot, inch, and pace were once based upon dimensions associated with the human body. The need for refinement and objectivity showed the inadequacy of using inherently variable reference standards. Today the practice is to measure potency relative to an arbitrarily defined but constant reference standard.

Suppose that a particular *standard preparation*, S, produces a response in individual animals of a population, and that the magnitude of the response is dependent upon the dose of S in the sense that mean response at a dose is determined by the dose. Writing z as the dose of S and u as the response of an animal, this dose-response relation can be expressed by a regression function

$$U \equiv E(u|z) = F(z) . \qquad (I.2.1)$$

I shall assume $F(\ )$ to be a monotonic function of z, at least within the range of z to be used experimentally, and without loss of essential generality I can regard $F(\ )$ as an increasing function. Now suppose that we also have for consideration a *test preparation*, T, and that responses to S and T can be described by the two regression functions

$$\left.\begin{aligned} U_S &= F_S(z) , \\ U_T &= F_T(z) . \end{aligned}\right\} \tag{I.2.2}$$

Assay of T relative to S requires that there be a number ρ, the *relative potency*, such that 1 unit (by weight or volume) of T has the same potency as ρ units of S, and that this ratio defines all equivalent doses of S and T. If this be true, the response of any animal to a dose z_T of T must be exactly what it would be had the dose been ρz_T of S. The regression functions are therefore related by

$$F_T(z) = F_S(\rho z) \tag{I.2.3}$$

for all z. This equation, central to the whole of bioassay, represents *similarity of response* to the two preparations. Unless it is true, the potency of T relative to S cannot be expressed by a single value, and S and T must be essentially different in biological effect. Although proposals have been made for defining measures of relative potency under less stringent conditions, I shall limit myself to situations conforming to (I.2.3). The aim then is to estimate the parameter ρ.

Individual responses will vary about U_S or U_T. A simple formulation that often adequately describes reality is that u is normally distributed with constant variance, i.e. u is $N(U, \sigma^2)$. Since attention usually concentrates on the mean of a number of responses at each dose, the central limit theorem is a protection against non-normality, but procedures for other distributions can be developed if required. Constancy of variance is more open to question. As a matter of statistical experience, quite considerable variance heterogeneity can be ignored without seriously affecting the estimation of ρ. In Lecture II I shall discuss a situation in which σ^2 itself is functionally dependent upon U.

3. Estimation

The standard experiment for bioassay consists of administering selected doses of S and T to suitable subjects, and obtaining a value of u for each administration; u might be a measure of growth on a whole animal, the weight of a particular organ of the animal, a biochemical measurement on its blood, or the amount of contraction produced in a strip of animal tissue suspended in a solution corresponding to a specified dose of S or T. From the values of u, classified according to preparation, dose and any other constraint of design, must be estimated ρ and commonly also other parameters of the model.

I shall look particularly at one form of response curve, not only because it is both commonly applicable and outstandingly simple but also because it is basic to a much wider range of assay models. Suppose that

$$U_S = f(\alpha + \beta \log z) , \qquad (I.3.1)$$

where $f(\ )$ is a known monotonic function. Define the inverse function by

$$Y = f^{-1}(U) \quad . \tag{I.3.2}$$

Then introduce the *dose metameter*, a transformation of dose,

$$x = \log z \quad . \tag{I.3.3}$$

The two dose-response relations can now be written

$$Y_S = \alpha + \beta x \quad , \tag{I.3.4}$$

$$Y_T = \alpha + \beta \log \rho + \beta x \quad , \tag{I.3.5}$$

the equations of two parallel lines separated by a horizontal distance $\log \rho$.

Far from being a highly artificial set of possibilities, equations (I.3.1) - (I.3.5) in fact represent a good approximation to many practical assay situations. Indeed, experimental data often do not conflict with the further striking simplification of equation (I.3.2) to

$$Y = U \quad , \tag{I.3.6}$$

and a constant variance of u about U . Taken in conjunction with a normal distribution for u , this simplification means that equations (I.3.4), (I.3.5) can be estimated by elementary linear regression calculations, subject to the constraint that the two lines must be parallel. The y corresponding to a response u is known as the *response metameter*. Write the lines

$$\hat{Y}_S = \bar{Y}_S + b(x - \bar{x}_S) \quad , \tag{I.3.7}$$

$$\hat{Y}_T = \bar{Y}_T + b(x - \bar{x}_T) \quad , \tag{I.3.8}$$

with obvious notation for the means of x and y and the estimated regression coefficient. The horizontal distance is

$$M = \bar{x}_S - \bar{x}_T + (\bar{y}_T - \bar{y}_S)/b \quad ; \qquad (I.3.9)$$

ρ is then estimated by R , the antilogarithm of M .

After the introduction of equations (I.3.1), (I.3.2), equation (I.3.6) may seem an anticlimax. The greater generality is needed, but before further developing this family of *parallel line* assays, I must comment briefly on another simple situation. This is appropriate when equation (I.3.1) is replaced by

$$U = f(\alpha + \beta z) \quad . \qquad (I.3.10)$$

With equation (I.3.6), the previous form of argument leads to

$$Y_S = \alpha + \beta z \quad , \qquad (I.3.11)$$

$$Y_T = \alpha + \beta\rho z \quad . \qquad (I.3.12)$$

The computational procedure now is to estimate two straight lines constrained to intersect at $z = 0$, and from these

$$\hat{Y}_S = a + b_S z \quad , \qquad (I.3.13)$$

$$\hat{Y}_T = a + b_T z \quad . \qquad (I.3.14)$$

The estimator of ρ is now

$$R = b_T/b_S \quad . \qquad (I.3.15)$$

Such *slope ratio* assays also have important practical applications, but I shall say little about them.

4. Validity Tests

The calculation of equations (I.3.7), (I.3.8) in order to obtain the estimator M involves familiar processes of analysis of variance for additive linear models, and provides opportunity for testing some of the assumptions underlying the estimation. Unless the experimental techniques are long-established and the statistical properties of the regression curve are well-known, the assumptions of linearity, constancy of variance, and normality are open to question as follows:

(i) Components of the sum of squares between doses can be isolated so as to correspond to types of deviation from linearity. In particular, a single degree of freedom representing quadratic regression is a good indicator of most forms of non-linearity.

(ii) Usually the experiment will be conducted on a relatively small number of doses of each preparation, with several independent responses measured at each dose. Error mean squares can then be calculated at each dose and their homogeneity tested by Bartlett's χ^2. This test is known to be sensitive to disturbances other than variance heterogeneity (such as non-normality), and should not be regarded as an absolute criterion for an analysis that assumes constant variance. Other tests may be used to look specifically for a trend in variance as dose increases, for example comparing a mean square for all high doses with a mean square for all low doses. If the assay is well-designed, conclusions on potency will be little affected by moderate inequalities of variance; therefore only very strong indications demand action.

(iii) Although normality of distribution of individual responses, y, about the expectation, Y, for a dose is implicit in the analysis, departure from this is seldom a serious worry. All important inferences depend upon mean responses per dose and derived quantities, and the central limit theorem therefore protects the analysis. In bioassay, the number of responses per dose is usually small (between 5 and 20), so that a powerful test of normality is impossible unless information from several experiments can be combined. Various distribution-free techniques for estimating ρ have been proposed, but I question whether circumstances often justify their inherent clumsiness.

The above tests refer to the purely *statistical validity* of the analysis. Serious evidence of invalidity is to be followed by search for a new model, departing from the assumption that u is $N(U, \sigma^2)$ or that equation (I.3.6) applies; I discuss an example in Lecture II. Of even greater concern is evidence against the *fundamental validity* represented by (I.2.3), or by (I.3.4), (I.3.5) for the most important class of assays. If no objections to simple normal linear regression appear but the S and T regressions deviate significantly from parallelism, the fundamental requirement for a dilution assay is contradicted: T does not behave as though it were exactly like S except for an unknown difference of dilution or concentration, and therefore its potency relative to S cannot be expressed by a single number ρ. This can happen at an early stage of development of a new assay technique because an inappropriate standard has been chosen; it can happen also with a well-established technique because something has gone wrong with the particular experiment, such as an impurity contaminating T or a failure to control some feature of the environment. Fundamental invalidity is a matter for the investigator rather than for the statistician, but it should usually mean both rejection of the data in hand and re-examination of the experimental conditions.

5. Interval Estimation

As is usual in the practice of statistical estimation, a single numerical estimator is inadequate; the scientific investigator wants to be able to say with some certainty that the true value of ρ lies between calculated limits. If he has concluded that the assay is valid, he will judge its quality largely by the narrowness of the interval between the limits.

Equation (I.3.9) shows M to be, apart from the additive constant $(\bar{x}_S - \bar{x}_T)$, the ratio of two linear functions of the observations. Under the assumptions governing the analysis, both $(\bar{y}_T - \bar{y}_S)$ and b are normally distributed, but their ratio certainly is not normal. A general theorem on interval estimation applicable here was fully stated first by Fieller (1944), though forms of it had previously been in use. Without regard to previous specific notation, suppose that a, b are two linear functions of a set of normally distributed observations, such that

$$\left.\begin{aligned} E(a) &= \alpha , \\ E(b) &= \beta , \end{aligned}\right\} \qquad (I.5.1)$$

and

$$\left.\begin{aligned} Var(a) &= \sigma^2 v_{11} , \\ Var(b) &= \sigma^2 v_{22} , \\ Cov(a,b) &= \sigma^2 v_{12} , \end{aligned}\right\} \qquad (I.5.2)$$

where

$$\mu = \alpha/\beta \qquad (I.5.3)$$

is to be estimated by

$$m = a/b \qquad (I.5.4)$$

and the v_{ij} are functions of the design matrix of the observations (that is to say of replication numbers and of independent variates). Suppose further that s^2, an unbiased estimator of σ^2, has been obtained as a mean square with f degrees of freedom in an analysis of variance, and that t is the deviate from the t-distribution with f degrees of freedom corresponding to a selected probability P (such as 0.95). Then with probability P

$$\left| \frac{(a - \alpha) - \mu(b - \beta)}{[s^2(v_{11} - 2\mu v_{12} + \mu^2 v_{22})]^{\frac{1}{2}}} \right| \leq t . \qquad (I.5.5)$$

Using (I.5.3), this inequality can be written

$$(a - \mu b)^2 \leq t^2 s^2 (v_{11} - 2\mu v_{12} + \mu^2 v_{22}) . \qquad (I.5.6)$$

Taking the equality sign and solving the quadratic for μ, two roots are obtained, m_L and m_U. The values of μ that satisfy (I.5.6) are usually all those for which

$$m_L \leq \mu \leq m_U . \qquad (I.5.7)$$

If

$$b^2 \leq t^2 s^2 v_{22} , \qquad (I.5.8)$$

the limits are exclusive rather than inclusive; the statement then to be made is that

$$\mu \leq m_L \text{ or } \mu \geq m_U . \qquad (I.5.9)$$

The possibility of exclusive limits is at first sight a strange feature of the theory; it enters logically, and its acceptance makes possible statements of limits with probability P for the ratio α/β .

The quantity

$$g = t^2 s^2 v_{22}/b^2 \qquad \text{(I.5.10)}$$

determines whether or not the limits will be inclusive. If b, the denominator of m, is large compared with its standard error, g will be much less than 1.0 ; conversely, if b differs from zero by an amount not much greater than its standard error, the limits can be exclusive. The limits, obtained from the equality sign in (I.5.6), can be written

$$m_L, m_U = \left[m - \frac{gv_{12}}{v_{22}} \pm \frac{ts}{b} \left\{ v_{11} - 2mv_{12} + m^2 v_{22} - g(v_{11} - \frac{v_{12}^2}{v_{22}}) \right\}^{\frac{1}{2}} \right] / (1-g). \qquad \text{(I.5.11)}$$

Complicated though this looks, it becomes very simple when g is small enough to be neglected (see Lecture III).

Fieller's theorem can be applied directly to the ratio $(\bar{y}_T - \bar{y}_S)/b$ in equation (I.3.9); here v_{12} is zero unless complexities of design produce a covariance. For example, in the simplest of assays with N_S, N_T subjects completely randomized for S and T,

$$v_{11} = \frac{1}{N_S} + \frac{1}{N_T},$$

$$v_{22} = 1/\{\Sigma(x - \bar{x}_S)^2 + \Sigma(x - \bar{x}_T)^2\},$$

$$v_{12} = 0.$$

These remain true for the important class of randomized block designs; incomplete blocks, complete blocks with some observations accidentally lost, and other complexities of design will usually alter all of v_{11}, v_{22} and v_{12}. The dose difference $(\bar{x}_S - \bar{x}_T)$ is added to each limit, and the antilogarithms then give the corresponding limits for the potency ρ . For slope ratio assays, the theorem is applicable to the ratio b_T/b_S in equation (I.3.15). In Lecture III, I shall be concerned with generalizations of the theorem.

6. Experimental Design

Consideration of (I.5.11) aids the planning of an assay so as to shorten the interval $(m_U - m_L)$. A strict mathematical minimization is not possible as a consequence of choices open to the investigator. He can, however, plan the wise use of a fixed total number of subjects (or total number of measured responses) by such steps as:

(i) Divide the subjects equally between S and T ;

(ii) Use any advance information on ρ so as to choose doses for which

$$\log \rho - (\bar{x}_S - \bar{x}_T)$$

is close to zero;

(iii) Space doses as widely as possible, subject to avoiding extremes at which the regression of y (or u) on x becomes markedly non-linear.

Further improvements in the precision of estimation may be obtained by choosing an experimental design with good combinatorial properties. Randomized blocks and Latin squares enable constraints of resources (e.g. grouping of animals into litters) or of execution (e.g. limits on work that can be done by one man in one day) to be incorporated into the design. Incomplete block designs, especially confounded factorials, provide possibilities for dealing with limitations of numbers in groups. If each subject can be used successively with several doses, cross-over designs can make precision depend only on intra-subject variation. All these devices are aimed at reducing the relevant σ^2 and its estimator, s^2.

I do not propose to discuss these topics in any detail.

7. Non-linearity

With this background, more general dose-response curves can be discussed. At any dose, z, of S, observed responses will have a frequency distribution, the density function (pdf) for which must depend in some way upon z if the experimental procedure is to be capable of exploitation for potency estimation. If a model can be proposed for this pdf, in terms of any number of parameters, the argument leading to equation (I.2.3) can be extended to show that substitution of ρz for z must give the pdf for T. One can scarcely hope that an explicit estimator of ρ with optimal properties of consistency, efficiency, and even sufficiency will exist. Provided that the experiment is reasonably large, however, maximization of the likelihood will be a practicable method of estimating ρ and all other parameters simultaneously, with a reasonable hope that the asymptotic optimality will not be negated by the smallness of the sample. Modern computational facilities make practicable the numerical maximization of such a likelihood function even if it involves as many as ten parameters.

In practice, data from a biological experiment are seldom adequate for estimating more than about four parameters intrinsic to a response curve. Since the response curve itself is only of subsidiary interest in the course of estimating ρ, an empirical model with few parameters can be more useful than a model with greater biomathematical exactness that demands many parameters.

Commonly u can be assumed normally distributed about U ; even if this is not true of the u actually measured, a simple power transformation (such as replacing u by $u^{\frac{1}{2}}$, u^{-1}, or log u) often gives approximate normality, and I now assume this to be incorporated in the definition of u . The variance of u ,

$$\mathrm{Var}(u|U) = E[(u - U)^2] = V(U) , \qquad (I.7.1)$$

may itself be a function of U , though quite often the preliminary transformation that normalizes u also makes its variance nearly constant. Note particularly that the variance function also implicitly introduces z and ρz by way of equations (I.2.2), (I.2.3). Maximum likelihood is then almost equivalent to estimation by weighted least squares, that is to say by minimizing

$$\Sigma \frac{(u - U)^2}{V(U)} , \qquad (I.7.2)$$

where summation is over all doses of S and T with $F(z)$ and $F(\rho z)$ substituted appropriately for U . The log likelihood introduces also the term $\Sigma\log[V(U)]$, which is less sensitive to changes in the parameters than is the expression (I.7.2). As is well known from general theory, the two optimizations are asymptotically the same (cf. §II.3) Again numerical optimization by computer is a practicable procedure. Experience indicates that the important parameters are those in $F(\ )$. Any additional parameters in the variance function are unlikely to be estimated satisfactorily from one assay, but pooled information from many can be adequate to determine the essential features of $V(\ )$.

The more general model still can often employ a partially linear representation. For suppose that, as in (I.3.1) ,

$$U = f(\alpha + \beta \log z) \qquad (I.7.3)$$

where the function f( ) *may* contain other parameters, but two essential conditions are that every parameter except α must be the same for both S and T and that z does not enter in any way other than in the expression $(\alpha + \beta \log z)$. Then the standard argument still shows that

$$\log \rho = (\alpha_T - \alpha_S)/\beta . \qquad (I.7.4)$$

Typically, additional parameters not explicitly stated in (I.7.4) may relate to asymptotes of the response curve. No regression relation between observed response and dose can be linear over an indefinite range of dose; a common type is the sigmoid curve, approximately linear over a moderate range, but approaching horizontal asymptotes as log z becomes very small or very large.

When parameters other than α,β are known, or can be given provisional values, the transformation to Y defined by (I.3.2) becomes possible. In terms of Y , the essential features of the regression are then represented in the linear form

$$\left.\begin{aligned} Y_S &= \alpha_S + \beta x , \\ Y_T &= \alpha_T + \beta x , \end{aligned}\right\} \qquad (I.7.5)$$

from which estimation of ρ proceeds much as in §3 . Of course, if (I.7.3) were to contain terms in z or $(\log z)^2$ or the like, this linearity disappears. Despite the restriction in (I.7.3) and the associated conditions, the general model still appears to provide a remarkably flexible system for empirical approximation to response curves, whether or not it is adequate for a description consistent with eventual theory.

In Lecture IV, I shall explore an important application of these ideas and comment on the iterative optimization of the likelihood, as well as mentioning alternative forms of (I.7.3). The linearizing transformation is not essential, convenient though it often is; equation (I.2.3) remains an absolute requirement for all that I discuss.

8. An Example

I end with a specimen of computer output from a recently developed general program for parallel line assays. This program is an improved version of one described by McArthur *et al*. (1966). It allows the user a choice of response metameters and is applicable to completely randomized and randomized block designs; it has greater output flexibility than the earlier program. The "Activators" are arbitrarily defined integers that control a wide range of options in output; in routine bioassay, one would seldom demand as much output as the example shows. I shall not discuss either the program or the example here, as the most important features should be self-explanatory. The enthusiastic reader can test his understanding of the principles by recalculation from the data.

The graphical representation has been organized so as to superpose the regression lines for S and T (points indicated by + and 1). Although this sacrifices the possibility of measuring M on the graph, it gives better visual representation of the evidence for any consistent deviation from linearity.

```
 ---------------------------------------------------
 :    DATA FOR    - BIOASSAY USING PARLIN 5   :
 ---------------------------------------------------


SOURCE OF DATA:       BRITISH STANDARDS INSTITUTION   (COD-LIVER OIL)
NAME OF ASSAY:        VITAMIN D3, BY CHICK BONE-ASH
DATE OF ASSAY:          1940
DATE OF ANALYSIS:     11/09/75, AT  8:58:09

C O M M E N T:           COMPLETELY RANDOMIZED PARALLEL LINE ASSAY

     ** FOR CURRENT DATA, THIS IS RUN NO.  1 **


          TABLE OF DATA (AS INITIALLY ENTERED)
          -------------------------------------

PREPARATION   0:     VITAMIN D3
            DOSE NO.  1  =      5.7600;    9 RESPONSES
      33.5000        33.0000        32.4000        33.7000        32.8000
      37.3000        33.1000        32.1000        29.5000
            DOSE NO.  2  =      9.6000;   10 RESPONSES
      36.2000        36.7000        39.5000        36.2000        35.4000
      35.6000        34.8000        37.0000        39.4000        34.2000
            DOSE NO.  3  =     16.0000;    8 RESPONSES
      41.6000        40.5000        39.1000        39.4000        43.0000
      37.9000        42.0000        42.4000

PREPARATION   1:     COD-LIVER OIL
            DOSE NO.  4  =     32.4000;    7 RESPONSES
      32.0000        33.9000        31.6000        32.7000        28.8000
      30.2000        33.1000
            DOSE NO.  5  =     54.0000;    6 RESPONSES
      32.6000        36.0000        34.8000        29.2000        34.6000
      37.7000
            DOSE NO.  6  =     90.0000;    8 RESPONSES
      35.7000        38.9000        40.3000        42.9000        43.9000
      42.8000        38.9000        38.6000
            DOSE NO.  7  =    150.0000;    7 RESPONSES
      44.0000        43.3000        44.2000        41.8000        43.7000
      38.4000        40.1000


 ---------------------------------------------------
 : METAMETERS FOR - BIOASSAY USING PARLIN 5   :
 ---------------------------------------------------


SOURCE OF DATA:       BRITISH STANDARDS INSTITUTION   (COD-LIVER OIL)
NAME OF ASSAY:        VITAMIN D3, BY CHICK BONE-ASH
DATE OF ASSAY:          1940
DATE OF ANALYSIS:     11/09/75, AT  8:58:10

C O M M E N T:           COMPLETELY RANDOMIZED PARALLEL LINE ASSAY

     ** FOR CURRENT DATA, THIS IS RUN NO.  1 **
```

```
        TABLE OF METAMETERS FOR ANALYSIS
        --------------------------------

NO TRANSFORMATION OF RESPONSE

PREPARATION    0:    VITAMIN D3
     DOSE METAMETER NO.  1  =  0.7604;    9 RESPONSES
     DOSE METAMETER NO.  2  =  0.9823;   10 RESPONSES
     DOSE METAMETER NO.  3  =  1.2041;    8 RESPONSES

PREPARATION    1:    COD-LIVER OIL
     DOSE METAMETER NO.  4  =  1.5105;    7 RESPONSES
     DOSE METAMETER NO.  5  =  1.7324;    6 RESPONSES
     DOSE METAMETER NO.  6  =  1.9542;    8 RESPONSES
     DOSE METAMETER NO.  7  =  2.1761;    7 RESPONSES

NO RESPONSE METAMETERS HAVE BEEN DELETED
    ------------------------------------------------
    :  RESULTS FROM  - BIOASSAY USING PARLIN 5  :
    ------------------------------------------------

SOURCE OF DATA:      BRITISH STANDARDS INSTITUTION    (COD-LIVER OIL)
NAME OF ASSAY:       VITAMIN D3, BY CHICK BONE-ASH
DATE OF ASSAY:         1940
DATE OF ANALYSIS:    11/09/75, AT  8:58:10

C O M M E N T:          COMPLETELY RANDOMIZED PARALLEL LINE ASSAY

    ** FOR CURRENT DATA, THIS IS RUN NO.  1 **


 ACTIVATORS USED WERE:
     IA1 = 3, IA2 = 0, IA3 = 0, IA4 = 1, IA5 = 3, IA6 =  3, IA7 = 0


                      MINIMUM RESPONSE FOR ANALYSIS =    28.799988
                      MAXIMUM RESPONSE FOR ANALYSIS =    44.199997

 EXAMINATION OF VARIANCE HETEROGENEITY
     BARTLETT KI-SQUARED =   3.6073, WITH    6 D.F.

     AVERAGE VARIANCE FOR HIGHEST DOSES =      4.039032, WITH   13 D.F.
     AVERAGE VARIANCE FOR LOWEST DOSES  =      3.658523, WITH   14 D.F.

|| RESIDUAL/ERROR MEAN SQUARES      =    4.42 WITH  1, 48 D.F.

 REGRESSION COEFFICIENT:     B =    17.026242,  VARIANCE =    1.869042
 EFFECTIVE SD PER RESP.: S/B  =     0.1284

 INDEX OF REGRESSION SIGNIFICANCE: T**2*V(B)/B**2
     FOR PROBABILITY 0.95, G =  0.0261
     FOR PROBABILITY 0.99, G =  0.0464


P O T E N C Y   E S T I M A T E S   A N D   L I M I T S
--------------------------------------------------------

PREPARATION    POTENCY          LIMITS(0.95)          LIMITS(0.99)

 COD-LIVER OIL
     1       0.1460         0.1243    0.1723       0.1177    0.1828
```

: ANALYSIS FOR - BIOASSAY USING PARLIN 5 :

SOURCE OF DATA: BRITISH STANDARDS INSTITUTION (COD-LIVER OIL)
NAME OF ASSAY: VITAMIN D3, BY CHICK BONE-ASH
DATE OF ASSAY: 1940

** FOR CURRENT DATA, THIS IS RUN NO. 1 **

A N A L Y S I S O F V A R I A N C E

SOURCE OF VARIATION	D.F.	SUM OF SQUARES	MEAN SQUARE
PREPARATIONS	1	6.870793	6.870793
REGRESSION	1	740.878351	740.878351
PARALLELISM	1	0.092872	0.092872
ALL QUADRATICS	2	1.985206	0.992603
RESIDUAL	1	21.133962	21.133962
DOSES	6	770.961185	
ERROR	48	229.281745	4.776703
T O T A L	54	1000.242930	

PREPARATION MEANS, FOR USE IN PLOTTING

PREPARATIONS STANDARD LISTED FIRST		MEAN OF X	MEAN OF Y	YBAR-B*XBAR
0	VITAMIN D3	0.9741	36.6037	20.0192
1	COD-LIVER OIL	1.8512	37.3107	5.7910

DOSE MEANS AND VARIANCES

(ONLY FOR DEVIATIONS EXCEEDING 0.3 TIMES S.E.)

SE	RESPONSES	X	Y-BAR	EXPTN	VARIANCE	DEVN	S.E.
STANDARD:	VITAMIN D3						
2	10	0.982	36.500	36.744	3.1200	-0.244	0.548
3	8	1.204	40.737	40.521	3.2570	0.217	0.567
PREPARATION 1	COD-LIVER OIL						
4	7	1.511	31.757	31.510	3.0895	0.247	0.543
5	6	1.732	34.150	35.287	8.7110	-1.137	0.774
6	8	1.954	40.250	39.064	7.7028	1.186	0.638
7	7	2.176	42.214	42.842	4.9514	-0.627	0.561

```
-----------------------------------------------------
:  DIAGRAM FOR   - BIOASSAY USING PARLIN 5   :
-----------------------------------------------------

SOURCE OF DATA:      BRITISH STANDARDS INSTITUTION   (COD-LIVER OIL)
NAME OF ASSAY:       VITAMIN D3, BY CHICK BONE-ASH
DATE OF ASSAY:          1940
   ** FOR CURRENT DATA, THIS IS RUN NO.  1 **
```

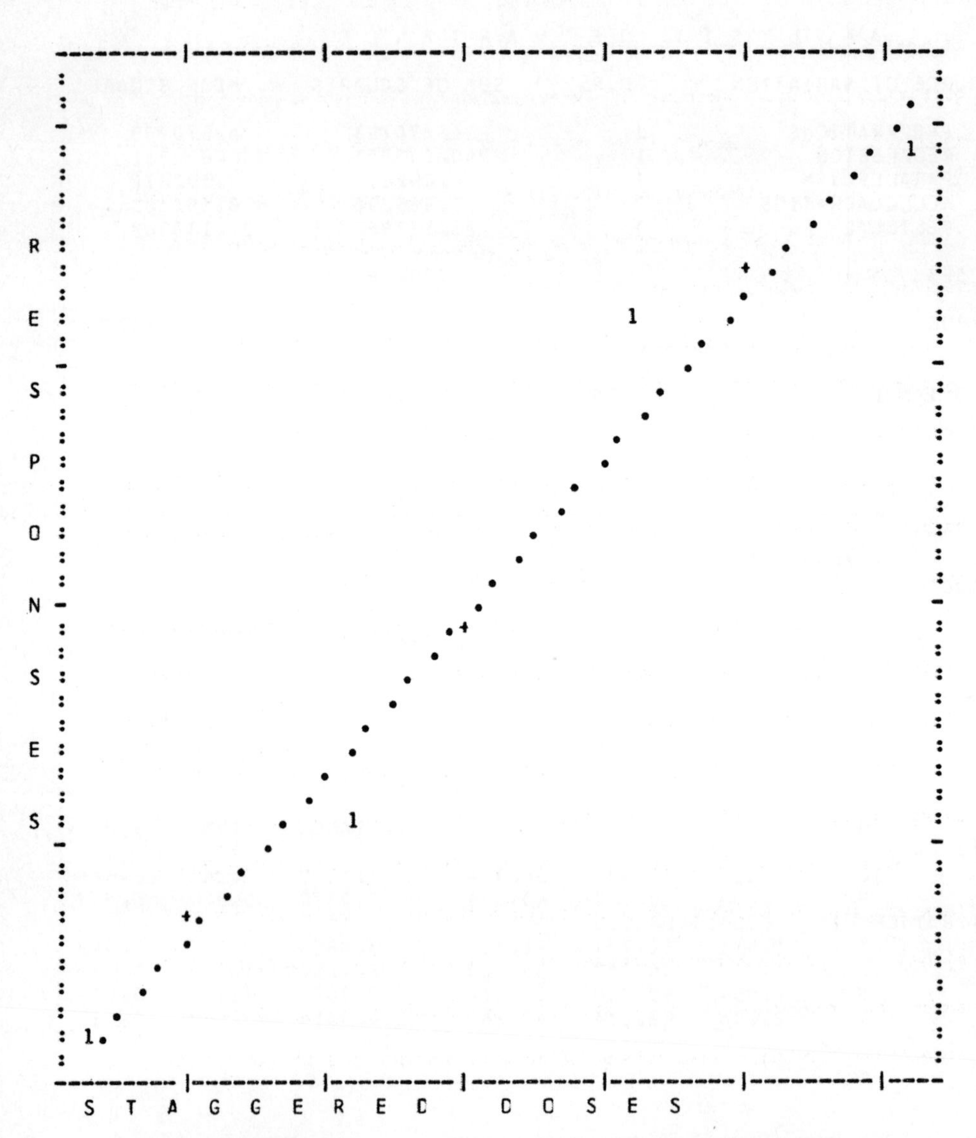


```
1 UNIT HORIZONTALLY =  0.0111 UNITS OF DOSE METAMETER
1 UNIT VERTICALLY   =  0.2833 UNITS OF RESPONSE METAMETER
```

BIOASSAY

Lecture II: RADIOLIGAND ASSAYS

1. The Techniques

Many assay methods have been developed in which antigenicity is estimated with the aid of radioisotopes. They require very small quantities of preparations. In *radioimmunoassays* antigens are labelled with radioisotopes, whereas in *immunoradiometric* assays antibodies are labelled; from the statistical viewpoint, the two types are similar. The literature is immense; Rodbard and Frazier (1974) and Rodbard and Hutt (1974) provide useful condensed accounts and references to other recent work of statistical interest.

Early users of these radioligand assays were so surprised by the smoothness of response curves that could be drawn freehand, and by the apparent precision of potency estimates, that they often neglected requirements of assay validity (§I.4) and the numerical assessment of precision. The temptation to rely upon linear interpolation or rough graphical procedures was perhaps encouraged because the typical response curves encountered were not well represented by the types of function that could be easily fitted by familiar statistical procedures. Assay practitioners did not have access to, or were not accustomed to using, powerful computing facilities and were perhaps intimidated by the labour of using unsophisticated desk calculators for non-linear optimization.

The response in these assays is a radiation count. In radioimmunoassays, to which I shall restrict discussion, the count made is either 'B' (on the bound fraction) or 'F' (on the free fraction). At zero dose, the bound count is usually symbolized by B_o, and at an effectively infinite dose (which is an experimental possibility) this count becomes 'N', the non-specific value. Different workers have advocated using as the response metameter at a dose:

$$\frac{B-N}{B+F-N} \quad \text{("fraction bound")}, \qquad \text{(II.1.1)}$$

$$\frac{F}{B+F-N} \quad \text{("fraction free")}, \qquad \text{(II.1.2)}$$

$$\frac{B-N}{B_o-N} \quad \text{("bound/zero-dose ratio")}, \qquad \text{(II.1.3)}$$

$$\frac{B-N}{F} \quad \text{("bound/free ratio")}, \qquad \text{(II.1.4)}$$

and reciprocals of these. Plotted against log dose (x), any of them may show an approximately linear regression at intermediate doses, and thus provide a basis for a graphical or simplified arithmetical assessment of $M = \log R$ as the horizontal distance between two lines. At extremes of dose, the asymptotic behaviour of the counts must be evidenced by non-linearity. I shall not attempt to present biological and physico-chemical details of the techniques.

2. Sigmoid Response Curves

The next stage in the development of an adequate statistical approach was the recognition that curvature in the dose-response relation might be taken into account by adopting a suitable class of regression function. Several workers in radioimmunoassay (I shall not discuss priorities here) found curves of the logistic type useful. The logistic function,

$$\frac{1}{1+e^{-(a+bx)}}, \qquad \text{(II.2.1)}$$

ranges from 0 at zero dose (x large and negative) to 1 at large doses, or the reverse if $b < 0$. Although variants on this and other sigmoids could be adapted to several of the functions of B and F mentioned above, chief attention has been given to (II.1.3). Various lines of development can be seen in papers by Ekins and Newman (1970), Healy (1972), Midgley *et al.* (1969), and Rodbard and Cooper (1970). In particular, the latter used a form of estimation involving a linear regression of

$$\log\{(B-N)/(B_0-B)\} \qquad \text{(II.2.2)}$$

on x to give (a+bx). In this and subsequent papers, Rodbard and his colleagues recognized the need for weighting the regression calculations so as to take account cf unequal precision at different parts of the scale.

The logistic function arises in mathematical models for many biological phenomena, though it can seldom be regarded as more than a first-order approximation to a reality that is much more complicated; frequently, biological theory cannot adequately explain the phenomena fully in terms of differential equations that would lead to a model involving many parameters. Although justification for using the logistic function in radioimmunoassay may be little more than empirical, it does seem to be quite widely applicable. This implies no belief that all radioligand assays can be similarly analyzed. Moreover, data more extensive than any ordinarily available would be needed in order to discriminate empirically between the logistic and alternative sigmoids such as

$$\int_{-\infty}^{a+bx} \frac{1}{\sqrt{2\pi}} \exp(-\tfrac{1}{2}t^2)\,dt \ ; \qquad \text{(II.2.3)}$$

no theoretical arguments for (II.2.3) have been advanced, but almost certainly it would be equally successful in fitting data.

3. Full Parameterization

The symbolism current among users of radioimmunoassay is not always internally consistent and is not ideal for a general statistical discussion. For the present purpose, therefore, I shall start again so as to maintain consistency with Lecture I. This does not affect the principles, or the outcome in terms of potency estimation. In particular, what I present here is very close to the more recent publication of Rodbard and his colleagues, notably Rodbard and Hutt (1974).

Although examination of functions such as (II.1.1) - (II.1.4) is helpful in exploring the nature of a dose-response curve, the fact that B_o and N must themselves be estimated adds to difficulties in any search for optimal estimators; it can also obscure the distinction between parameters and their numerical estimates. A generally better procedure in most problems of applied statistics is to concentrate attention on the independently observed and measured quantities (here the individual counts at all doses), and to express these in terms of a model.

As in Lecture I, I shall write

$$U \equiv E(u|z) = F(z) \qquad \text{(II.3.1)}$$

for the expectation of a single response (count) at dose z. In line with what has been suggested as a logistic model, suppose that

$$U = C + \frac{D-C}{1+e^{-2Y}}, \qquad \text{(II.3.2)}$$

where

$$Y = \alpha + \beta x, \qquad \text{(II.3.3)}$$

x as usual being $\log z$. In radioimmunoassays, β is negative. On the x scale, as dose becomes very small U approaches the upper asymptote D, the parameter corresponding to the previous B_o; at large doses, U approaches the lower asymptote C, the parameter corresponding to the previous N. The other two parameters, α and β, relate to the linear representation discussed in Lecture I. If C and D were known, the parametric transformation

$$Y = \frac{1}{2} \log \left(\frac{U-C}{D-U}\right), \qquad \text{(II.3.4)}$$

in which Y is defined as the *logit* function of (U-C)/(D-C), would lead to the linear representation of the response curve as (I.3.4) exactly as in § I.3. Neither the factor 2 in (II.3.2) nor the corresponding factor ½ in (II.3.4) is essential; many people choose to omit these, but they have some advantages (§ II.5). In practice the four parameters α, β, C, D must be estimated simultaneously.

Next the behaviour of individual counts must be considered. The simplest hypothesis would suggest that at any dose u would have a Poisson distribution with mean U. Since U is usually greater than 100 and may be much greater than 1000, such a distribution could well be approximated by a normal distribution, though working with the Poisson itself would present no difficulties. Empirical evidence cannot be denied; the variance of u about U is usually greater than a Poisson distribution would show, presumably because of the accumulation of variation from many other sources. Rodbard and Cooper (1970) and Rodbard (1971) have discussed the components of variance thoroughly, but they end with models which they themselves recognize to be too complicated for use; not surprisingly, they conclude that a normal distribution will suffice, provided that the variance itself is taken to be a function of U that increases with U rather more rapidly than for a Poisson.

Since u is normally distributed, the likelihood function, L, for the totality of counts can be specified. In fact, with $V(U)$ as in (I.7.1),

$$\log L = \text{const} - \frac{1}{2}\sum \log[V(U)] - \frac{1}{2}\sum \left[u - C - \frac{D - C}{1 + \exp\{-2(\alpha+\beta x)\}}\right]^2 / V(U) , \tag{II.3.5}$$

where summation is over every count and its corresponding dose. Maximizing the likelihood is almost equivalent to minimizing the weighted sum of squares of $(u-U)$, the expression (I.7.2). As is well known, maximum likelihood and minimum χ^2 estimation are asymptotically identical, and it matters little which is used. With $V(U)$ as in (II.4.2) below, the additional term modifying the weighted sum of squares or χ^2 is simply a multiple of $\sum \log U$; once a reasonable approach to the optimal parameter estimates for U has been made, the geometric mean of U values will be very stable under small numerical changes in the estimates.

4. The Variance Function

Rodbard and Hutt (1974) found that a quadratic function of the expected response metameter represented variance satisfactorily for various functions of u of the kind listed in §II.1, despite the fact that all of these cannot be mutually consistent. As applied to u itself, the quadratic must vanish at $U = 0$, and so may be written

$$V(U) = HU(1 + JU) . \tag{II.4.1}$$

Though I agree that this function is adequate, I find the alternative

$$V(U) = H\,U^I \tag{II.4.2}$$

more adaptable and more easily manipulated. In each, H is of little intrinsic interest, since only relative variances at different doses affect the estimation process. Thus (II.4.1) can behave like the Poisson variance for small J, but can approach proportionality to U^2 for large J. In theory, (II.4.2) can show a wider range of patterns of behaviour, but $1.0 \leq I \leq 2.0$ appears to include those that are practically useful. One objection to (II.4.1) may be that data could chance to indicate negative values for J and H, so making $V(U)$ implausible at small U ! In practice, the two functions can behave so similarly that data do not readily show either as the better.

Although the parameter J in (II.4.1) or I in (II.4.2) could be estimated at the same time as all other parameters (i.e. β, C, D and an α for each preparation in an assay), this is not usually desirable. A typical assay includes many preparations, with replicate counts at each of one or more doses. Most of the information on $V(U)$ comes from large numbers of separate mean squares, with perhaps 1,2 or 3 degrees of freedom, and their associated values of $\bar{u}$.

The estimation of a variance function of specified form, relating $E(s^2)$ to U from the evidence provided by many values of s^2 and $\bar{u}$ and the simple assumptions of normality is a problem that does not seem to have been formally discussed, though to set up a maximum likelihood procedure for it is not difficult. For the one large set of data to which I have had access, providing several hundred values of $\bar{u}$ and mean squares, a plot of s^2 against $\bar{u}$ showed a clear tendency for s^2 to increase with $\bar{u}$ but a tremendous scatter of points. It was very obvious that many alternatives within the families (II.4.1) and (II.4.2) would fit. A plot of log s^2 against log $\bar{u}$ did not contradict the linearity appropriate to (II.4.2); indeed it did not rule out either $I = 1.0$ or $I = 2.0$, though an intermediate slope seemed more appropriate. Hence a compromise of $I = 1.5$ seemed worth consideration.

This picture from extensive data is some indication that for estimating I a single assay will be inadequate. Moreover, applied statistics has produced many examples to show the rather small effects of varying the relative weighting of data provided that account is taken of major factors. Consequently, a wiser policy seems to be to assess the variance parameter J or I from the accumulated evidence of many assays, and, while being prepared to modify this in the light of further data, to use it as though it were a known quantity. Modern computational facilities allow occasional checks on the robustness of the procedure to be made by trying alternative functions on either side of that being used. Section II.6 illustrates the procedure, and confirms that the consequences of quite large differences in the choice of I may be small for the potency estimates. Possibly a theoretical study of the magnitude of effects of errors in the variance function could reinforce or modify this recommendation.

5. The Optimization

The nature of radioimmunoassay and its uses are such that it is common practice to include many test preparations in the same experiment. Several doses of the standard preparation will be tested, with replicate counts at each, and also replicates at zero and 'infinite' doses to improve information on D and C ; each test preparation (perhaps 50 or more) may be represented by replicate counts at only one or two doses. As should be clear from Lecture I, each preparation will have its own α , but β, C, D must be common to the whole experiment. On the Y or logit scale, the dose-response relations become a series of parallel lines, and, as in (I.3.9), (I.7.4),

$$M = (\hat{\alpha}_T - \hat{\alpha}_S)/\hat{\beta} \tag{II.5.1}$$

estimates $\log \rho$.

How is the maximization of $\log L$ in (II.3.5) to be achieved, with $V(U)$ now replaced perhaps by $HU^{1.5}$? Until a few years ago, the standard method for all such problems was to expand the left hand sides of the equations

$$\frac{\partial}{\partial\theta} (\log L) = 0 , \tag{II.5.2}$$

formed for every parameter θ , to the next order of terms so as to have linear equations with coefficients

$$\frac{\partial^2}{\partial\theta_1 \partial\theta_2} (\log L) \tag{II.5.3}$$

evaluated at an approximation to the desired set of $\hat{\theta}$. In a manner illustrated in many books and papers that discuss numerical determination of maximum likelihood estimates, the linear equations provide adjustments that improve the approximate $\hat{\theta}$, and with the revised values the process is repeated (e.g. Finney, 1971). Thus the true $\hat{\theta}$ to satisfy (II.5.2) are approached iteratively (I ignore here the possibility of multiple maxima for L). Moreover, when the $\hat{\theta}$ have been obtained, the inverse of the Hessian matrix, with elements

$$\frac{\partial^2}{\partial\theta_1 \partial\theta_2} (\log L)$$

evaluated at $\hat{\theta}$, is the negative of the matrix of variances and covariances for the parameters. Consequently, Fieller's theorem can be applied to (II.5.1) in the standard way to give limits for the true $\log \rho$, except for the dependence upon asymptotic theory.

This process used to be very tedious. It was also exposed to risk of algebraic slips, because it required that all first and second differential coefficients of $\log L$ with respect to every parameter be formed and evaluated numerically for each cycle of iteration; for a new mathematical form of $F(z)$ that may be wanted only occasionally, errors in signs and omissions of terms in the derivatives can easily occur. To-day the labour can be transferred to computers, with the possibility of many more iterations; if provision is made for steady narrowing of inter-vals, all the differential coefficients can be replaced by first and second-order differences. The early stages of approach to the optimum need not follow the classical method of iteration, but can instead use alternative routines that are more efficient for computers. I have found the procedure devised by Nelder and Mead (1965), with minor modifications of my own, very effective (in the form of my own program, MINISQUARE); furthermore, several excellent alternatives exist. The former terrors of the arithmetic are eliminated, and I can no longer see any excuse for using graphical or "quick-and-dirty" procedures of the kind that at one time rightly were valued.

One point has been neglected, the factor H in (II.4.2), though doubtless the steps relating to it are obvious to those familiar with parameter estimation for truly linear models. The residual weighted sum of squares of deviations, the minimal value obtained by minimization, is asymptotically a χ^2 if the model is adequate; it can be written (cf. II.3.5)

$$\sum \left(u - \hat{C} - \frac{\hat{D} - \hat{C}}{1+\exp\{-2(\hat{\alpha}+\hat{\beta}x)\}} \right)^2 / HU^I . \qquad (II.5.4)$$

By equating this quantity to its expectation, that is to say its degrees of freedom or the number of independent counts less $(k + 3)$, where k is the number of preparations, H can be estimated. If some or all doses have replicate counts, the sum of squares in (II.5.4) can be analyzed between and within doses, so as to give two mean squares whose ratio tests whether deviations from the model are large relative to true replication.

I have done no more than outline the kind of program required; any computer centre that commonly handles data relating to statistical estimation should have a general program that can readily be modified to deal with a particular parametric model, and the details need not occupy us further.

One reason for preferring to retain the factor 2 in (II.3.2) is that a close similarity is then maintained with the alternative metametric transformation corresponding to (II.2.3), namely

$$U = C + (D - C) \int_{-\infty}^{Y} \frac{1}{\sqrt{2\pi}} e^{-\frac{1}{2}t^2} dt . \qquad (II.5.5)$$

The two models will not lead to identical values of $\hat{\alpha}$, $\hat{\beta}$, since the parameters have different mathematical meaning. Nevertheless, unless the data are very peculiar, iteration for either model can begin from the same initial approximations. A reasonable concordance of all parameter estimates, and indeed comparability at intermediate stages of iteration, is then a useful verification that the potency estimates are robust to variations in the exact assumptions made about response curves.

6. An Example

Dr. S. Cekan has kindly allowed me to use data from a series of assays executed according to the methods of Brenner *et al.* (1973). I have looked in particular at an assay of oestradiol preparations. Four counts were made at each of 6 doses of S, 4 more under non-specific or infinite dose conditions, and 8 at zero dose. The 8 counts averaged 1680, the non-specific averaged 38, and the others fell between these values. Many test preparations were included, but only at one dose of each (with duplicate counts).

The behaviour of potency estimates depends rather much upon how the counts for a preparation lie relative to the extremes of count found for S. Table 1 summarizes the results for 3 preparations, obtained by minimum χ^2 estimation; it shows how little the estimate R is affected by choice of the variance power, I. The limits R_L, R_U have been calculated for a probability 0.95.

TABLE 1

Potency estimates for three oestradiol preparations

	Variance function has			
	I = 2.0	I = 1.5	I = 1.0	I = 0.0
Preparation with mean count 1064				
R	57	58	58	58
R_L	38	46	49	49
R_U	85	73	69	68
Preparation with mean count 535				
R	217	217	217	217
R_L	173	187	190	181
R_U	271	253	249	262
Preparation with mean count 51				
R	14000	13000	13000	11000
R_L	6000	5000	3000	45
R_U	32000	33000	52000	2230000

7. Experimental Design

A physical measuring instrument is of little use for comparing quite different stimuli which lie in a range to which the instrument is largely insensitive. Thus a balance constructed to distinguish differences in weight in the range 50g - 500g is unlikely to discriminate between 0.1mg and 1.0mg, and will be quite incapable of handling weights such as 100kg. The sigmoidal shape of many response curves used in bioassay has similar consequences. At very low or very high doses, difference in response is a poor and insensitive indicator of differences in the potency of doses applied. Consequently low and high doses of test preparations should be avoided (as the third section of Table 1 illustrates) because they give relatively little information - though pre-existing knowledge of a preparation may be so slight that the investigator cannot judge whether a dose is "low" or "high" , in which case a trial of several doses should be mandatory.

If the full analysis described above is to be conducted, reasonably good estimation of C and D is important. Therefore, for the benefit of the assay as a whole, some counts at a very high dose and others at zero or a very low dose are needed, but they should not be repeated for each preparation. In the light of these recommendations, Cekan's choice of doses and replications may well have been close to the best use of fixed total resources.

8. A Suggested Variation

The method proposed in this lecture has used a model in which u is normally distributed about U with variance proportional to U^I . As is easily seen from elementary consideration of the transformation of variates, to a first-order approximation $u^{1-I/2}$ will have a variance that is almost constant. In the absence of any positive evidence for the normality of u, one might examine a model which states that $u^{1-I/2}$ is normally distributed about $U^{1-I/2}$ with constant variance. Thus with $I = 1.5$ the transformation

$$u^* = u^{1/4} \tag{II.8.1}$$

might be tried for each count, with

$$U^* = U^{1/4} \tag{II.8.2}$$

where U is as previously defined. For $I = 2$, the appropriate formula is

$$U^* = \log U . \tag{II.8.3}$$

I do not know if this form of the model has been tried. It may be quite unsuitable, with perhaps marked skewness of distribution for u^*, but possibly it leads to a simpler computational routine.

BIOASSAY

LECTURE III: COMBINATION OF ASSAY RESULTS

1. The Problem

Some bioassays are undertaken once only for the particular test preparations, perhaps because materials are being screened with a view to selecting for further study those that show some evidence of useful potency or because new batches of a familiar drug must be checked for acceptability of quality. At other times, several assays contribute information on the potency of a particular test preparation relative to a fixed standard. One reason is that physical conditions in a laboratory prevent execution of a single assay large enough for the precision wanted and the experiment is therefore repeated on several occasions. Another reason is that from time to time supplies of a widely agreed reference standard near exhaustion and steps must be taken to replace it by a new standard; a cooperative study involving several groups of people may be undertaken in order to measure the potency of the new standard relative to the old. Publications of the World Health Organization include many reports on such collaborative studies.

Thus a particular relative potency, ρ , may be estimated by several independent experiments. How is a single estimate to be formed? In some circumstances, a very simple procedure is adequate, but any approach to exact general theory is difficult. I shall discuss one recent development that displays some interesting aspects of maximum likelihood theory. The full story has been presented in a series of papers (Armitage, 1970, Armitage and Bennett, 1974, Armitage, Bennett and Finney, 1976, Bennett, 1963).

2. Weighted Means

The commonest method of combining independent estimates of a parameter is by a weighted mean, using reciprocals of variances as weights. Unfortunately, at the heart of the bioassay techniques that I am discussing is a ratio of normally distributed quantities; this does not have a true variance. When the variance of responses is sufficiently small, however, an asymptotic variance can be used. More exactly, if g is small enough to be neglected, the limits for a ratio obtained as (I.5.11) can be written

$$m_L, m_U = m \pm \frac{ts}{b}(v_{11} - 2mv_{12} + m^2 v_{22})^{\frac{1}{2}} , \qquad \text{(III.2.1)}$$

which is just what would be obtained by regarding

$$\text{Var}(m) = s^2(v_{11} - 2mv_{12} + m^2 v_{22})/b^2 \qquad \text{(III.2.2)}$$

as an estimated variance for m and using the t-deviate to give the limits.

As I described in § I.3, a single parallel line assay leads to the estimate of log potency in (I.3.9), which can be written

$$M = \{\bar{y}_T - \bar{y}_S - b(\bar{x}_T - \bar{x}_S)\}/b ; \qquad \text{(III.2.3)}$$

this has the form of a ratio of two linear functions of the data (correlated because b appears in numerator and denominator); and therefore both Fieller's theorem and the approximation of (III.2.1), (III.2.2) apply. I need not write the formulae in detail. Suffice it to say that, if a series of assays gives M_i, $\text{Var}(M_i)$ $(i=1,2, \ldots, k)$, then a weighted mean is

$$\bar{M} = \Sigma W_i M_i / \Sigma W_i , \qquad \text{(III.2.4)}$$

where

$$W_i = 1/\mathrm{Var}(M_i) \qquad \text{(III.2.5)}$$

and

$$\mathrm{Var}(\bar{M}) = 1/\Sigma W_i \quad . \qquad \text{(III.2.6)}$$

If each variance has been based upon a mean square, s_i^2 , with many degrees of freedom, or if (as in § 3) all the s_i^2 can be pooled into a single mean square, this can be satisfactory (in the sense of approximating well to the method of the sections that follow).

Equation(III.2.2) is obtainable directly as a "variance" for a ratio by the crude method of squared differentials; this general method for assigning some sort of variance to an otherwise intractable function of observations is not easily located in standard textbooks but is an important last resort of statisticians! A particular merit here is that it enables results from totally different types of assay to be combined - assays using different experimental techniques and even (with a futher approximation) slope ratio assays together with parallel line ones. The theory now to be introduced is more limited in application though more soundly based. It is related to the theory of linear functional relations; Sprent's survey of this field (1966) includes a mention of the problem in slightly different form.

3. Formulation

Suppose that each of k parallel line assays estimates the same potency, ρ , and that for assay i the estimator of $\log \rho$ is

$$M_i = a_i/b_i \ , \qquad \text{(III.3.1)}$$

where a_i has the form of the numerator of (III.2.3) and (c.f. I.5.1, I.5.2)

$$\left.\begin{aligned} E(a_i) &= \alpha_i , \\ E(b_i) &= \beta_i , \end{aligned}\right\} \qquad (III.3.2)$$

and

$$\left.\begin{aligned} Var(a_i) &= \sigma_i^2 v_{i11} , \\ Var(b_i) &= \sigma_i^2 v_{i22} , \\ Cov(a_i, b_i) &= \sigma_i^2 v_{i12} . \end{aligned}\right\} \qquad (III.3.3)$$

There is no reason to suppose the α_i, β_i constant from assay to assay, but the condition that all assays estimate the same ρ means that

$$\mu = \alpha_i / \beta_i \qquad (III.3.4)$$

is constant. If all the assays used the same experimental technique, with the same species of animal and the same definition of response and good control of conditions, one might expect the variance per response to be almost constant. The theory that follows assumes

$$\sigma_i^2 = \sigma^2 \text{ for all } i , \qquad (III.3.5)$$

but is unlikely to be seriously misleading for assays of about equal size in which the σ_i^2 do not differ greatly. The limitation which this assumption introduces is serious; any departure from it would introduce all the complications of the Behrens distribution and its generalizations, together with all the controversies these have generated.

Each assay will produce its own error mean square, s_i^2 . Equation (III.3.5) justifies forming a pooled estimate, s^2 , which will usually have abundant degrees of freedom, f (even with $k = 3$, f is likely to be at least 30 and might be more than 60).

An analogous formulation will deal with slope ratio assays; I shall not describe it. This does not mean, however, that estimates from the two types of assay can be combined.

4. The Likelihood

Under a hypothesis H_0 that relaxes the condition (III.3.4), the log likelihood of the data can be written

$$L_0 = C(\sigma^2) - \frac{1}{2\sigma^2} \sum \frac{1}{v_{11}v_{22} - v_{12}^2} \{v_{22}(a - \alpha)^2 - 2v_{12}(a - \alpha)(b - \beta) + v_{11}(b - \beta)^2\} ; \quad \text{(III.4.1)}$$

here and subsequently all summations are over $i = 1,2, \ldots,k$, and, in order to avoid clumsy symbolism, the subscript i is omitted from v_{i11}, a_i, α_i etc. within summations. Here $C(\sigma^2)$ is a function of σ^2 not involving other parameters. Maximization of L_0 clearly leads to

$$\hat{\alpha}_i = a_i \ , \ \hat{\beta} = b_i \qquad \text{(III.4.2)}$$

corresponding to using each M_i as an estimator of a μ_i .

Two hypotheses logically nested within H_0 impose constraints on the μ_i :

$$H_{k-1} : \mu_i = \mu \text{ , to be estimated;} \qquad \text{(III.4.3)}$$

$$H_k : \mu_i = \theta \text{ , a specified value.} \qquad \text{(III.4.4)}$$

Looking first at H_k , the log likelihood is seen to be

$$L_k = C(\sigma^2) - \frac{1}{2\sigma^2} \sum \frac{1}{v_{11}v_{22} - v_{12}^2} \{v_{22}(a - \beta\theta)^2 - 2v_{12}(a - \beta\theta)(b - \beta) + v_{11}(b - \beta)^2\}, \quad \text{(III.4.5)}$$

which is maximized by

$$\left.\begin{aligned}\hat{\beta}_i &= \frac{\theta(v_{i22}a_i - v_{i12}b_i) - (v_{i12}a_i - v_{i11}b_i)}{v_{i22}\theta^2 - 2v_{i12}\theta + v_{i11}},\\ \hat{\alpha}_i &= \theta\hat{\beta}_i,\end{aligned}\right\} \quad \text{(III.4.6)}$$

again a simple explicit solution. For H_{k-1}, the log likelihood L_{k-1} is exactly as in (III.4.5) except that the unknown μ replaces the specified θ. Differentiating L_{k-1} with respect to μ gives another equation to be satisfied, which can be written

$$\hat{\mu} = \sum \left(\frac{\hat{\beta}\{v_{22}a - v_{12}(b - \hat{\beta})\}}{v_{11}v_{22} - v_{12}^2} \right) \Big/ \sum \left(\frac{v_{22}\hat{\beta}^2}{v_{11}v_{22} - v_{12}^2} \right). \quad \text{(III.4.7)}$$

Iteration with (III.4.7) and (III.4.6), after replacing θ by $\hat{\mu}$, will usually lead rapidly to a numerical solution.

I have not commented on the estimation of σ^2. Strict maximum likelihood procedure requires σ^2 to be estimated afresh with each maximizing of L, the log likelihood. Under H_0, maximizing with respect to the other parameters makes the second term of L_0 zero, and maximizing $C(\sigma^2)$ is essentially equivalent to using the pooled estimate s^2 for σ^2. Under H_{k-1} and H_k, additional information on σ^2 is available (corresponding with the test statistics Z_{k-1} and Z_1 below), but asymptotically this makes no difference and s^2 can be used for each. Because of the nesting of the hypotheses, so that H_0, H_{k-1}, H_k represent increasing levels of constraint with reductions of $(k - 1)$ and 1 in the numbers of free parameters,

$$Z_{k-1} = 2\,(L_{0,max} - L_{k-1,max}) \quad \text{(III.4.8)}$$

and

$$Z_1 = 2\,(L_{k-1,max} - L_{k,max}) \qquad \text{(III.4.9)}$$

will be asymptotically distributed as χ^2 with $(k - 1)$ and 1 d.f. respectively. A little algebra enables these to be rewritten as

$$Z_{k-1} = \frac{1}{s^2} \sum \frac{(a - b\hat{\mu})^2}{v_{22}\hat{\mu}^2 - 2v_{12}\hat{\mu} + v_{11}} \qquad \text{(III.4.10)}$$

and

$$Z_1 = \frac{1}{s^2} \sum \frac{(a - b\theta)^2}{v_{22}\theta^2 - 2v_{12}\theta + v_{11}} - Z_{k-1} \,. \qquad \text{(III.4.11)}$$

These useful forms suggest that,when the number of degrees of freedom for s^2 is not particularly large,the distributions may be better approximated by regarding $Z_{k-1}/(k - 1)$ and Z_1 as variance ratios with $(k - 1)$, f and 1, f degrees of freedom respectively.

What do the tests mean? Evidently Z_{k-1} is appropriate to testing the restricted hypothesis that all assays share the same μ as compared with the unrestricted hypothesis that permits a separate μ_i to each assay. This is essentially a test of validity of the hypothesis on which the assays rest, namely that a single relative potency measures the relation between the two preparations despite possible differences in the α_i and β_i. The *test* of Z_1 is less directly valuable except for the rare situation of theory presenting a value θ of μ as being especially interesting. Nevertheless Z_1 is useful for interval estimation. By extending the ideas of Fieller's theorem (§ I.5), we can assign an interval to μ which consists of those values of θ that are not rejected by a significance test using Z_1.

5. The Performance of Potency Estimates

When the separate values of a_i/b_i are not very different, the procedures I have described obtain the estimate $\hat{\mu}$ under H_{k-1} and determine probability limits for it without difficulty. Apart from the extra degrees of freedom contained in Z_{k-1} and available for testing the discrepancies between the k estimates, the general appearance of the method is not very different from an ordinary application of Fieller's theorem. Wide differences between the separate values of a_i/b_i can alter the situation substantially. Although the more bizarre consequences seem unlikely to occur unless Z_{k-1} is so large as to invite rejection of the hypothesis of a single μ, a study of possibilities is interesting.

The method of estimating μ can be considered in three ways: it is a maximization of likelihood, a minimization of χ^2, and a minimization of a weighted sum of squares for a non-linear model. All methods depend upon minimizing the function

$$J(\mu) = \sum \frac{(b\mu - a)^2}{v_{22}\mu^2 - 2v_{12}\mu + v_{11}} . \qquad \text{(III.5.1)}$$

As μ becomes large (positive or negative), $J(\mu)$ approaches

$$J_\infty = \Sigma b^2/v_{22} ; \qquad \text{(III.5.2)}$$

necessarily $v_{22} > 0$, so that $J_\infty > 0$ except for the trivial case of all $b_i = 0$. Thus a graph of $J(\mu)$ against μ shows an approach to a horizontal asymptote at both ends. Expanding $J(\mu)$ in inverse powers of μ gives

$$J(\mu) = J_\infty - 2\mu^{-1} \sum \left(\frac{ab}{v_{22}} - \frac{b^2 v_{12}}{v_{22}^2}\right) + \ldots \qquad \text{(III.5.3)}$$

In general, therefore, the asymptote will be approached from opposite sides at the two ends. If the second term in (III.5.3) is zero, the nature of the approach will depend upon higher order terms and can be one-sided, but the necessary condition seems not to embody any interesting special case.

Commonly $J(\mu)$ has a single minimum at $\hat{\mu}$, and of course

$$Z_{k-1} = J(\hat{\mu})/s^2 . \qquad (III.5.4)$$

For $k=1$, the minimum of $J(\mu)$ is at zero, and the curve for $J(\mu)$ necessarily has this one minimum together with one maximum above the asymptote - a well-known way of looking at Fieller's theorem. For $k>1$, the minimum is not zero unless all a_i/b_i are exactly equal, and the extent to which $J(\hat{\mu})$ exceeds zero is at the heart of the test of the hypothesis of a common μ. Moreover

$$Z_1 = [J(\theta) - J(\hat{\mu})]/s^2 . \qquad (III.5.5)$$

The function $J(\mu)$ can have as many as k real minima; I know of no proof that it cannot have more, though I do not think this possible*. Of course the minima will be separated by up to $(k - 1)$ maxima, with an additional maximum beyond the minima at one end of the scale. None of the denominators in $J(\mu)$ can vanish for real μ, and therefore the function is always finite. One can easily construct an artificial example with k minima, by taking the curve for $k = 1$, a standard Fieller

---

* Elementary arguments show that at most (4k-2) extrema could occur. During the seminar, Dr. W. Knight suggested to me a simple demonstration of the possibility of up to (2k-1) minima separated by (2k-2) maxima with an additional maximum outside the range of the minima.

curve, sliding it horizontally until the new minimum occurs where the original curve was close to its asymptote, repeating the sliding process $(k - 1)$ times and then writing $J(\mu)$ to correspond to adding the ordinates for the k curves. Thus each minimum will be associated with one of the a_i/b_i, though it will not necessarily occur very close to that ratio. In practice, values of the a_i, b_i are commonly such that several minima coalesce and eliminate the intervening maxima. Within this framework, the range of possibilities is immense. There can be one very small minimum and others at substantially larger values of $J(\mu)$, even with some minima above the asymptote; there can be two or more minima with nearly or exactly equal values of the function. The maxima may be low hills separating one minimum from the next or tall pinnacles reaching far above the asymptote.

Note that, if σ^2 were known, $-2J(\mu)$ would be the log-likelihood; the form of $J(\mu)$ is independent of the value of σ^2 or s^2, and even with only an estimator of σ^2 present there is evidently a relation to the likelihood. The number of minima is not determined by the absolute values of the v-factors; multiple minima can occur even when $Var(a_i)$, $Var(b_i)$ are all large. Evidently the ratios $v_{i11}: v_{i12}: v_{i22}$, and their relations to a_i/b_i, are what matter, but I know of no simple way of discovering the number of minima nor even of any criterion for the existence of a single minimum. In elementary statistical problems, and certainly in those with only a single parameter, likelihood functions with multiple maxima are uncommon (except for trivial instances where probabilities depend only on a single function of the parameter such as $\theta(1 - \theta)$ or θ^2). Barnett (1966) and Edwards (1972, §8.2) have commented on other examples.

What are the implications for the estimate of μ . The global minimum of $J(\mu)$, the *minimum minimorum*, is still to be regarded as $\hat{\mu}$. Use of Z_1 to determine an interval estimate is equivalent to drawing a horizontal line on the diagram at a height derived from (III.5.5), namely

$$J(\hat{\mu}) + \sigma^2\chi^2_{[1]}$$

or

$$J(\hat{\mu}) + s^2F(1,f) ,$$

where χ^2 or F is read from a table at 0.95 or other appropriate probability. Values of μ for which the line lies above the $J(\mu)$ curve are "acceptable"; the remainder are rejected. Thus instead of a simple interval one may now have a set of values of μ consisting of several disjoint segments, one possibly including infinity. Whether the more unusual patterns occur at "interesting" probability levels like 0.95 or 0.99 or only when limits are found for a probability such as 0.02 will depend upon σ^2 or s^2 . The geometric behaviour of $J(\mu)$ depends solely upon the numerical magnitudes of the a_i, b_i and the replications and choices of dose manifested in the ratios $v_{i11}: v_{i12}: v_{i22}$; whatever the geometry permits can occur when limits are found for any level of probability if σ^2 or s^2 is of the right magnitude.

My impression is that the more bizarre possibilities are of theoretical but not practical interest. Given any $J(\mu)$, with several distinct minima, I believe that a value of σ^2 small enough to make more than one minimum influence a statement about probability limits for μ will usually also be so small as to cause Z_{k-1} to represent very convincing evidence against accepting all the assays as estimating a single μ. A proof of this conjecture, or a further investigation of it, would be interesting.

6. An Example

A simple example with $k = 2$ can illustrate many of the possibilities. I do not suggest that the numerical values used in it correspond to any genuine bioassay results. Suppose that the two assays have the same (known) variance σ^2 per response and that other values are:

i	a_i	b_i	v_{i11}	v_{i12}	v_{i22}
1	$-d$	1	1	0	p
2	d	1	p	0	1

with $p \geq 0$ and $d \geq 0$. Evidently for separate estimates

$$\left.\begin{aligned} \hat{\mu}_1 &= -d \,, \\ \hat{\mu}_2 &= d \,. \end{aligned}\right\} \qquad \text{(III.6.1)}$$

If the method of §III.2 were applicable, the asymptotic variances

$$\left.\begin{aligned} \mathrm{Var}(\hat{\mu}_1) &= \sigma^2(1 + pd^2) \,, \\ \mathrm{Var}(\hat{\mu}_2) &= \sigma^2(d^2 + p) \,, \end{aligned}\right\} \qquad \text{(III.6.2)}$$

lead to the weighted mean

$$\bar{\mu} = d \left(\frac{d^2 - 1}{d^2 + 1}\right)\left(\frac{p - 1}{p + 1}\right) \,, \qquad \text{(III.6.3)}$$

with

$$\mathrm{Var}(\bar{\mu}) = \sigma^2 \, \frac{(1 + pd^2)\,(d^2 + p)}{(d^2 + 1)\,(p + 1)} \,. \qquad \text{(III.6.4)}$$

Note that necessarily

$$-d < \bar{\mu} < d \,. \qquad \text{(III.6.5)}$$

For this situation

$$J(\mu) = \frac{(\mu + d)^2}{p\mu^2 + 1} + \frac{(\mu - d)^2}{\mu^2 + p} , \qquad \text{(III.6.6)}$$

with an asymptote at

$$J_\infty = (1 + p)/p . \qquad \text{(III.6.7)}$$

A very strange situation occurs if $d = 1$ and $p = 1$; then

$$J(\mu) = 2 ,$$

a constant for all μ . Hence $\hat{\mu}$ is indeterminate, and at any chosen probability all values of μ are equally acceptable, whereas (III.6.3) and (III.6.4) suggest the approximation

$$\bar{\mu} = 0 \pm \sigma .$$

If d is unspecified but $p = 1$,

$$J(\mu) = 2(1 + \frac{d^2 - 1}{\mu^2 + 1}) ; \qquad \text{(III.6.8)}$$

in this case, the curve lies entirely above the asymptote if $d > 1$ and so leads to an infinite value for $\hat{\mu}$. On the other hand, if $d < 1$, $J(\mu)$ is always below the asymptote, with a minimum at $\hat{\mu} = 0$. We easily verify from Z_{k-1} that

$$\chi^2_{[1]} = 2d^2/\sigma^2 \qquad \text{(III.6.9)}$$

is to be used as a test of agreement between the assays, this being equivalent here to the simple test of

$$\alpha_1 = \alpha_2 .$$

From Z_1 ,

$$\chi^2_{[1]} = \frac{2(1 - d^2)}{\sigma^2} \cdot \frac{\theta^2}{\theta^2 + 1} \qquad \text{(III.6.10)}$$

tests whether θ is an acceptable value for μ. If G is the tabular value for a chosen probability level of $\chi^2_{[1]}$, the corresponding limits for μ are $\pm\theta$, where

$$\theta^2 = \frac{G\sigma^2}{2 - 2d^2 - G\sigma^2} \quad ; \qquad \text{(III.6.11)}$$

if the denominator of this expression is negative the limits should be taken to be infinite. On the other hand, uncritical use of the weighted mean, from (III.6.3), (III.6.4), would correctly give $\bar{\mu} = 0$ but would set the limits at

$$\pm\, \sigma[G(1 + d^2)/2]^{\frac{1}{2}} \quad ;$$

for $d < 1$ these limits are closer to zero than $\pm\theta$. This corresponds to underestimating θ^2 by an absolute amount equal to

$$\theta^2[1 - \frac{(2 - 2d^2 - G\sigma^2)\ (1 + d^2)}{2}] \ . \qquad \text{(III.6.12)}$$

Thus the proportional error in estimating θ^2 from simple weighting approaches 100% as $(2d^2 + G\sigma^2)$ approaches 2.

Despite the earlier disclaimer, the example is perhaps not as unrealistic as might be imagined. The same proportions could be preserved in two assays of simple design. For example, an assay with r subjects at each of 2 doses of each preparation, the doses being unit distance apart on the logarithmic scale, is easily seen to have

$$v_{11} = v_{22} = 1/r\ , \quad v_{12} = 0\ . \qquad \text{(III.6.13)}$$

If two such assays happened to have

$$a_1 = d,\ a_2 = -d,\ b_1 = b_2 = b\ , \qquad \text{(III.6.14)}$$

the essential features of the artificial example I have described would apply. The simple weighted mean method gives

$$\bar{\mu} = 0 \ , \tag{III.6.15}$$

$$\mathrm{Var}(\bar{\mu}) = \sigma^2(1 + \frac{d^2}{b^2})/2rb^2 \ . \tag{III.6.16}$$

If $d > b$, $\hat{\mu}$ is infinite. If $d < b$, the test for agreement between the two assays, based on Z_{k-1}, takes the form

$$\chi^2_{[1]} = 2rd^2/\sigma^2 \tag{III.6.17}$$

The statistic Z_1 to be used in determining limits becomes

$$\chi^2_{[1]} = \frac{2r\theta^2(b^2 - d^2)}{\sigma^2(\theta^2 + 1)} \ . \tag{III.6.18}$$

The relation to (III.6.9) , (III.6.10) is obvious.

Now we would scarcely accept the two assays as being in satisfactory agreement unless the test statistic in (III.6.17) were less than 5.0, and therefore

$$d^2 < 2.5\sigma^2/r \ .$$

Any single assay in which b is less than 5 times its standard error is of seriously reduced value because the biological system has been poorly responsive to dose changes, and therefore one should have

$$b^2 > 25\sigma^2/r \ .$$

For limits at probability 0.95, $G = 3.84$, and use of $\bar{\mu}$ instead of the likelihood method represents an erroneous reduction in θ^2 by about 10% at worst. If b were to be as small as 3 times its standard error in each assay, the underestimation of θ^2 by using the variance of a weighted mean to give limits could be 35%; if b were only twice its standard error, the weighted mean calculations would continue to give inclusive limits whereas the likelihood calculations would give exclusive limits. Replacement of σ^2 by an estimated s^2 makes the position somewhat worse. Of course, all these comparisons assume underlying normal error distributions.

One may reasonably conclude that when good assays are being combined - good in the sense that each gives a precise estimate of its own regression coefficient very different from zero - simple weighted means will perform well. The danger lies in combining assays that are not individually good. Further study of the example in this section, without limitation to $p = 1$, reveals the possibility not only that $J(\mu)$ has two minima but even that $\hat{\mu}$ falls well outside the interval $(-d, d)$. If this can happen when $k = 2$, far more worrying things can happen for larger k, as the study of many artificial examples with $k = 3$ has verified.

BIOASSAY

LECTURE IV: THE NEED FOR CONTEMPORARY COMPARISONS

1. The Problem

Though I have not explicitly discussed the matter, I have implied that the usual practice in bioassay is for the standard and test preparations to be used within a single experiment. The types of bioassay that I am discussing are such that T behaves as though it were a dilution (or a concentration) of S, with any diluent involved being inactive in respect of responses. Consequently, as in (I.2.3), a dose z of T must produce responses exactly the same as would a dose ρz of S. Implicit in this statement is the requirement that doses of S and T shall be used under identical conditions, or that, where conditions necessarily vary from one dose to another, the allocation shall be strictly randomized.

For some assay techniques, the size of an experiment, or the amount of labour required or the numbers of animals necessary as subjects, would make convenient the arrangement of work on S on one day and that on T on another. The alternative of using both S and T on each day, but with half the replication each time, is less attractive because it is more troublesome. In a laboratory concerned with the routine assay of new batches of a material, an even more important consideration is that contemporary testing of S and T involves expending half the total experimental effort on the standard. Not surprisingly, investigators wish to avoid the apparent wastage of time and materials on S, the preparation whose properties are so well known that it is used as a reference standard.

The professional statistician will usually advise against

estimating a dose-response regression on one day and using that regression for calculating the relative potencies of test preparations on other days. Even though the day for S is selected at random, so as to eliminate bias, the responsiveness of subjects may vary substantially from day to day. If this happens, the horizontal distance between regression lines, and consequently the estimate of potency, will be influenced by an additional component of variance. Moreover, if S and T have been tested on only one day each, the day-to-day variation cannot be estimated; indeed, existence of such variation may not be suspected, and ρ may simply be seriously over- or under-estimated because of the chance occurrence of a large day-to-day difference. Even if the assay were extended so that several days were devoted to S and several more to T, the information might scarcely be adequate for estimating the appropriate variance component, and in any case the deleterious effects on the precision of R would persist.

Statistical caution naturally urges that the dangers be avoided by ensuring that potency is estimated from intra-day or intra-experiment comparisons between S and T. Is this caution necessary? The great numbers of assays that have used a fixed standard with changing test preparations, conducted by pharmaceutical companies and governmental agencies, must contain much information on day-to-day variation in the response regression for S, but published evidence is surprisingly scarce. Moreover the findings for one type of assay or one laboratory may have no general applicability. I recently encountered one extensive series of data that illustrates well the logical problem and methods of studying it, even though the series is, in some respects, atypical. I propose to discuss these briefly, but need first to look at another form of dose-response relation.

2. Quantal Responses

In some dose-response situations, no measurement of response on a continuous scale is practicable; all that can be observed is a dichotomy: a subject responds or fails to respond, and no intermediate state is recognizable. Commonly the response involved is death, a natural choice in assessments of insecticidal potency, but others such as the production of convulsions in mice or the cure of a pathological condition can be appropriate. If n subjects all receive the same dose, and this dose is such that the probability of response is P, the probability that r responses occur (in circumstances that ensure independent reactions in the subjects) is

$$\binom{n}{r} P^r (1 - P)^{n-r} \quad . \qquad \text{(IV.2.1)}$$

Suppose that for the standard preparation the population of subjects has a probability density function $f(x)$ to describe the frequency distribution of individual tolerances of S as measured on the logarithmic scale. The subjects that respond to a dose whose logarithm is x will be all whose tolerances are less than x, that is to say $P(x)$ where

$$P(x) = \int_{-\infty}^{x} f(t) \ dt \ . \qquad \text{(IV.2.2)}$$

Unless the population contains subjects that manifest the characteristic response without administration of S (so that $P(x) \geq C > 0$ as $x \to -\infty$) or subjects that are totally immune to S (so that $P(x) \leq D < 1$ as

$x \to \infty$), $P(x)$ must change monotonically from 0 to 1 as x increases from $-\infty$ to ∞.

The model most commonly adopted is perhaps the normal distribution of log tolerances (Finney, 1971),

$$f(x) = \frac{1}{\sigma\sqrt{2\pi}} \exp\left[-(x-\mu)^2/2\sigma^2\right] , \qquad \text{(IV.2.3)}$$

which can be reparameterized to correspond with (I.3.1) as

$$f(x) = \frac{\beta}{\sqrt{2\pi}} \exp\left[-(\alpha+\beta x)^2/2\right] . \qquad \text{(IV.2.4)}$$

Some have advocated the alternative

$$f(x) = \frac{1}{2}\,\mathrm{sech}^2(\alpha+\beta x) , \qquad \text{(IV.2.5)}$$

which leads to the logistic function

$$P(x) = \frac{1}{1+e^{-2(\alpha+\beta x)}} . \qquad \text{(IV.2.6)}$$

The comparison with the sigmoid curves suggested in Lecture II for radioimmunoassays will be noted. There the logistic model is perhaps more reasonable in relation to the biological and chemical reactions involved, whereas for quantal assays the normal distribution of log tolerances is usually preferred. Empirical studies do little to help, since only an experiment with many doses and many subjects at each dose could distinguish clearly between the goodness of fit for (IV.2.4) and for (IV.2.5). Indeed other sigmoid curves with even less theoretical basis have been examined and are rarely rejected by the evidence of good single experiments. This indeterminacy in respect of mathematical models is, of course, only what would be found in other forms of statistical analysis if alternatives not very different from the normal distribution were to be tried. For bio-assay purposes, it is a protection rather than a worry, since it shows that conclusions obtained by assuming (IV.2.4) will remain essentially correct even if some other form of $f(t)$ is, in reality, nearer to the truth.

When several doses of a single preparation have been tested, with r_i responses out of n_i at dose i, the joint log likelihood of the observations can be written as

$$L = \text{constant} + \Sigma r_i \log[P(x_i)] + \Sigma(n_i - r_i)\log[1 - P(x_i)] \;, \qquad \text{(IV.2.7)}$$

where possibly we are taking

$$P(x) = \int_{-\infty}^{\alpha+\beta x} \frac{1}{\sqrt{2\pi}} e^{-\frac{1}{2}t^2} dt \;, \qquad \text{(IV.2.8)}$$

by simple rearrangement of (IV.2.2) and (IV.2.4). If Y is defined as the normal equivalent deviate (NED) corresponding to a probability P, so that

$$P = \int_{-\infty}^{Y} \frac{1}{\sqrt{2\pi}} e^{-\frac{1}{2}t^2} dt \qquad \text{(IV.2.9)}$$

provides a unique relation between Y and P, then

$$Y = \alpha + \beta x \qquad \text{(IV.2.10)}$$

is a linearization of the dose-response relation analogous to that described in §I.3. For data from two preparations, equations (I.3.4), (I.3.5) hold, and the parameters can be estimated by maximizing L.

Many years ago, the quantity $(Y + 5.0)$ was defined as the *probit* of P, as a device for avoiding the frequent occurrence of negative numbers for analyses that had to be performed with rather primitive calculators. Standard Newton-Raphson procedures for optimization were rearranged to give a convenient iterative system of calculation for maximizing L . If

$$\left.\begin{aligned} Y_S &= a_S + bx \ , \\ Y_T &= a_T + bx \ , \end{aligned}\right\} \tag{IV.2.11}$$

are taken as provisional equations for S and T , with a_S, a_T, b reasonable numerical guesses, then for each dose one can calculate

$$y = Y + (p - P)/Z \ , \tag{IV.2.12}$$

$$W = \frac{nZ^2}{P(1 - P)} \ , \tag{IV.2.13}$$

where

$$p = r/n \ , \tag{IV.2.14}$$

Y is taken from (IV.2.11), P is determined by (IV.2.9) for Y , and

$$Z = \frac{1}{\sqrt{2\pi}} e^{-\frac{1}{2}Y^2} \ . \tag{IV.2.15}$$

This stage is aided by the existence of tables that give Y as a function of P and also Z and W/n as functions of Y . Then the calculation of a weighted linear regression of y on x , with W as the weight and with the constraint that the fitted regression lines shall be parallel, gives new versions of (IV.2.11). The process is repeated until two successive versions of (IV.2.11) agree closely. The final coefficients are maximum likelihood estimators of the parameters; as in (I.3.9) or (II.5.1) ,

$$M = (\hat{\alpha}_T - \hat{\alpha}_S)/\hat{\beta} \tag{IV.2.16}$$

estimates $\log \rho$. Moreover, general maximum likelihood theory provides asymptotic variances for the numerator and denominator of M from quantities used in the regression calculations:

$$\left.\begin{aligned} \text{Var}(\hat{\alpha}_T - \hat{\alpha}_S) &= (\sum_S W)^{-1} + (\sum_T W)^{-1} , \\ \text{Var}(\hat{\beta}) &= [\sum_S W(x - \bar{x}_S)^2 + \sum_T W(x - \bar{x}_T)^2]^{-1} \\ \text{Cov}(\hat{\alpha}_T - \hat{\alpha}_S, \hat{\beta}) &= -(\bar{x}_T - \bar{x}_S) \text{ Var}(\hat{\beta}) , \end{aligned}\right\} \qquad \text{(IV.2.17)}$$

where $\sum_S, \sum_T$ denote summations over doses of the two preparations. With these, Fieller's theorem can be applied.

Details of the derivation of the formulae have often been published (Finney, 1971). The method is readily adapted to alternative forms of $f(t)$. Indeed, at one time considerable controversy existed about whether or not equation (IV.2.6) was a better model to take than equation (IV.2.8), and the availability of methods of estimation which were simpler in their arithmetic requirements than the true maximization of likelihood was sometimes advanced in favour of (IV.2.6). Today we can appreciate that very large bodies of data would be necessary in order to demonstrate empirically the superiority of one of the two models, and good computational facilities eliminate both the labour of full maximization of L and the temptation to use simplified arithmetic that will sometimes run into trouble. The choice of model, therefore, becomes little more than a matter of philosophical outlook, of little consequence for any practical conclusions drawn. A set of data for which the two models lead to appreciably different conclusions regarding potency should immediately be suspected and scanned for anomalous features.

3. Alternative Computations

The computations outlined in §2 were excellent for calculators (and operators) well adapted to the repetitive calculation of linear regressions. The flexibility of the modern computer makes other methods practicable. Many subroutines exist for maximizing an arbitrary mathematical function without requiring, at every step, the evaluation of first and second derivatives of the function implicit in the earlier method. Systematic evaluation of the function itself on a carefully chosen set of parameter-values, and the use of differences for reducing the intervals explored in the parameter spaces as the maximum is approached, can be more rapid and more easily generalized. I have a program called BLISS (named after the originator of the method of §2) that requires very little input for routine use, that provides several choices of function for $f(x)$ in order to check the robustness of conclusions, that incorporates a generalized model (§IV.4), and that outputs estimates of relative potency and other relevant statistics without requiring any supplementary calculation. Though I have a natural fondness for my own program, I imagine that others at least as satisfactory exist. Once $\hat{\alpha}_S$, $\hat{\alpha}_T$, $\hat{\beta}$ have been obtained, potency estimation proceeds by equation (IV.2.16)

BLISS first requires the user to supply some conventionalized instructions on the model to be adopted, the number of iterations required, and the various output options; the latter include not merely a listing of the data but also information on the progress of iteration that is occasionally useful in detecting why convergence was unsuccessful and in indicating how a new run can be improved. The data are then read and stored. The program next allows the user to specify, for each parameter, a guess at the estimate together with a step length to be used initially in exploring alternative values; these are the starting point for iteration. Alternatively, the program will make a preliminary scan of the data from which it determines its own starting point.

Iteration begins by evaluating L at the vertices of a simplex in the parameter space, defined by the starting values and the steps. Attention is shifted to the maximum of these values, step lengths are reduced, and the cycle is repeated. Various safeguards ensure that step lengths can be increased or decreased to prevent the process "sticking" too soon. Convergence is regularly checked against a criterion based on agreement between successive values of L. When adequate convergence has been achieved, a final phase is entered; first and second differential coefficients of L are estimated numerically from differences, and are used both to make a Newton-Raphson approach to the maximum and to give an inverse of the Hessian matrix as an asymptotic estimation of variances and covariances.

The output lists the parameter estimates and their standard errors, the variance-covariance matrix, and the estimate of ρ and its limits at several levels of probability. In addition extensive tabulation of L can be obtained, centred on the maximum value but with tabulation intervals as wide or as narrow as may be wished in respect of each parameter. A virtually unlimited number of test preparations can be included in the one analysis, though naturally the run time becomes much longer if k, the number of preparations, is large, since run time depends upon the calculations associated with a $(k+1) \times (k+1)$ matrix.

Although the program is primarily intended for the normal and logistic probability functions, (IV.2.8) and (IV.2.6), four others have been provided, all of course using the linear representation

$$Y = \alpha + \beta x .$$

They are

$$P(x) = \frac{1}{2}\left[1 + \frac{Y}{(1 + Y^2)^{\frac{1}{2}}}\right] , \qquad \text{(IV.3.1)}$$

$$P(x) = \frac{1}{2} + \pi^{-1} \tan^{-1} Y , \qquad \text{(IV.3.2)}$$

$$P(x) = \begin{cases} 0 & \text{if } Y < 0 , \\ \sin^2 Y & \text{if } 0 \le Y \le 90^{\circ} , \\ 1 & \text{if } Y > 90^{\circ} , \end{cases} \qquad \text{(IV.3.3)}$$

$$P(x) = \begin{cases} 0 & \text{if } Y < 0 , \\ Y & \text{if } 0 < Y \le 1 , \\ 1 & \text{if } Y > 1 . \end{cases} \qquad \text{(IV.3.4)}$$

These are not suggested for regular use, but they are occasionally worth trying as extremes to test the sensitivity of a potency estimate to rather violent changes in the model assumed for the data.

4. A Generalization of the Model

As I have hinted in §IV.2, the subjects may be such that a fraction C will respond however small the dose that is applied, and a fraction $(1 - D)$ will fail to respond however great the dose, where of course

$$0 \le C < D \le 1 .$$

Equation (IV.2.1) is unaffected, but in it P must now take the form

$$P = C + (D - C)\ P(x)\ , \qquad \text{(IV.4.1)}$$

where $P(x)$ is as in (IV.2.2) with any of the special cases for $f(x)$. This is clearly closely related to equation (II.3.2), but the binomial probability distribution for responses at a dose still leads to a log likelihood of the form of (IV.2.7) to be maximized instead of an expression like (II.3.5) to be minimized. Any of the methods of estimating parameters can be extended to include $\hat{C}$ and $\hat{D}$ (or to constrain either of these to be zero). As in Lecture II, $\hat{C}$, $\hat{D}$ make no direct contribution to potency estimation, though they will reduce its precision considerably if they are not close to 0, 1 respectively; equation (IV.2.16) remains unaltered.

5. Assay of Pertussis Vaccine

A technique known as the mouse protection test (Kendrick *et al.*, 1947) is often used for estimating the potencies of batches of pertussis vaccine. Mice are injected with selected doses of vaccine; 12 days later, each mouse is challenged with a standardized exposure to infection from a virulent strain of *Bordetella pertussis*. Those mice that survive for a further 15 days are regarded as responding to the vaccine, and quantal assay procedures are used for estimating the potency of a new batch of vaccine relative to a reference standard.

In the UK, manufacturers of pertussis vaccine are required to include the reference standard in every assay of new batches. Consequently, for the data I shall discuss, the possibility of economy by omission of the standard did not arise, but the availability of extensive data from routine assays raised the issue of whether the policy was ideal.

6. The Data

Almost two years ago, I was invited to comment on results for the standard vaccine, S , obtained in the 32 assays conducted at the National Institute for Biological Standards and Control in 1970; data for test vaccines in these assays do not enter into the discussion. In each assay, the same 3 doses of S were used, with 32 mice per dose. Table 6.1 shows the data. Preliminary scrutiny had suggested that variation from day to day might be entirely random, with the true response curve being the same throughout in a position well represented by the totals of 215, 531, and 850 responses out of 1024 mice. If this were so, even though simultaneous trial of S and T had to continue according to the Regulations, the precision of potency estimation might be improved by pooling the data on S from the previous 10 or 20 assays with that for the current assay, and perhaps later reducing the number of mice allocated to S in any one assay.

Though I was worried by one or two of the extremes, such as assays 6 and 19, my first impression was that quite possibly these were data for which the cautious outlook of §IV.1 was unnecessary. What follows is an adaptation of the publication that resulted from a fuller examination (Finney, Holt and Sheffield, 1975).

7. Maximum Likelihood Analysis

The question to be examined was whether the 32 sets of data in Table 6.1 could all be fitted by a single α and β . If they could not, natural alternatives would be to keep β constant but estimate 32 different α parameters or even to fit both α and β separately for each line.

TABLE 6.1

Numbers of mice surviving challenge with *Bordetella pertussis* after inoculation with the reference vaccine (batches of 32 inoculated)

Assay No.	Date in 1970	Dose of vaccine*		
		0.2	1.0	5.0
1	5/1	8	16	19
2	5/1	6	13	27
3	15/1	9	17	27
4	22/1	10	14	29
5	29/1	4	11	25
6	5/2	13	24	32
7	12/2	10	16	31
8	19/2	7	15	29
9	26/2	5	13	23
10	25/3	3	14	24
11	2/4	3	18	22
12	16/4	2	17	23
13	23/4	4	15	28
14	20/5	5	12	27
15	4/6	6	12	25
16	11/6	9	25	32
17	18/6	10	19	29
18	25/6	6	20	27
19	23/7	2	8	26
20	6/8	11	18	27
21	27/8	4	13	19
22	3/9	4	12	20
23	10/9	5	16	24
24	17/9	4	17	31
25	24/9	11	24	31
26	30/9	10	19	28
27	8/10	9	15	26
28	5/11	6	20	32
29	26/11	5	18	23
30	9/12	8	23	29
31	17/12	9	23	27
32	31/12	7	14	28
Total		215	531	850

* For arithmetical convenience, the unit of dose throughout this Lecture is 0.1 international units; the actual doses were 0.02, 0.10, 0.50 i. u.

If the 96 observations of $(n = 32, r)$ summarized in Table 6.1 are used and the two parameters α, β are estimated, the goodness of fit can be examined by the usual χ^2. For the normal tolerance distribution, this gives

$$\chi^2_{[94]} = 209.13 ,$$

suggestive of very considerable variability beyond that from the binomial distributions of responses at a dose. Estimation of 33 parameters simultaneously by maximum likelihood would have been an intimidating (though not impossible) task before the days of computers; BLISS encountered no difficulty in fitting $\alpha_1, \alpha_2, \ldots, \alpha_{32}$ and a single β, and gave a residual

$$\chi^2_{[63]} = 71.92 .$$

This clearly indicates a very satisfactory fit to the data, whereas the difference between the two values,

$$\chi^2_{[31]} = 137.21 ,$$

is a strong indicator of the need for separate α-parameters. In terms of the linearization represented by the NED transformation, the data conform excellently to 32 parallel lines, but these lines are markedly different in position.

The calculations were repeated with (IV.2.5), (IV.2.6) in place of (IV.2.4), (IV.2.8). Numerically, of course, the findings were different, but the conclusions were exactly the same; the 33 parameter model fits excellently, but the constraint of having a single α is unacceptable.

Rather than present the values of $\hat{\alpha}_i$, it is more in keeping with discussions of relative potency to look at the "50% immune doses", the estimated doses at which 50% of the mice would have survived the subsequent pertussis challenge. The logarithm of this so-called ImD50 (in other applications of the method known as the ED50 or 50% effective dose) is estimated for occasion i by

$$m_i = -\hat{\alpha}_i/\hat{\beta} \quad , \tag{IV.7.1}$$

the value of x for which $Y=0$. Fieller's theorem (I.5) can be applied to this ratio. Table 7.1 shows the estimate and its limit at probability 0.95 for each day of assay. In parallel columns, I have inserted corresponding figures from the logistic model to illustrate how little it matters in practice which is used.

The striking thing about Table 7.1 is the enormous difference in ImD50 on different occasions. The range is from 0.026 i.u. per mouse on occasion 6 to 0.208 i.u. per mouse on occasion 19; between February and July the reference standard vaccine has apparently declined eight-fold in potency! The ImD50 column shows no regular pattern, and a large difference can occur in assays only a week apart. There is no suggestion that a very few anomalous assays account for the heterogeneity or that some simple seasonal trend will explain most of the differences. On the contrary, the analysis confirms the fears of the cautious statistician that pertussis vaccine assays, like many others, will be untrustworthy unless based on contemporary comparisons between the S and T preparations.

The same form of analysis has been applied to comparable bodies of data for the years 1971, 1972 and 1973; all point even more strongly than 1970 to the existence of large and important differences between

TABLE 7.1

Estimates of ImD50 and limits at probability 0.95

Assay No.	Equation (IV.2.4)		Equation (IV.2.5)	
	ImD50	Limits	ImD50	Limits
1	1.34	0.82 - 2.17	1.35	0.82 - 2.23
2	1.11	0.68 - 1.80	1.13	0.69 - 1.85
3	0.74	0.46 - 1.20	0.74	0.45 - 1.21
4	0.73	0.45 - 1.18	0.74	0.45 - 1.21
5	1.60	0.98 - 2.61	1.63	0.99 - 2.68
6	0.26	0.16 - 0.44	0.26	0.16 - 0.45
7	0.57	0.35 - 0.92	0.58	0.35 - 0.95
8	0.82	0.51 - 1.34	0.83	0.51 - 1.37
9	1.51	0.93 - 2.46	1.53	0.93 - 2.52
10	1.53	0.94 - 2.50	1.53	0.93 - 2.52
11	1.38	0.85 - 2.24	1.35	0.82 - 2.23
12	1.46	0.90 - 2.39	1.44	0.88 - 2.37
13	1.06	0.65 - 1.72	1.06	0.65 - 1.74
14	1.25	0.77 - 2.04	1.27	0.78 - 2.09
15	1.32	0.82 - 2.16	1.35	0.82 - 2.23
16	0.32	0.19 - 0.53	0.32	0.19 - 0.55
17	0.55	0.33 - 0.89	0.54	0.33 - 0.90
18	0.75	0.46 - 1.22	0.74	0.45 - 1.21
19	2.08	1.27 - 3.41	2.09	1.26 - 3.47
20	0.62	0.38 - 1.02	0.62	0.37 - 1.01
21	2.03	1.24 - 3.34	2.09	1.26 - 3.47
22	2.03	1.24 - 3.34	2.09	1.26 - 3.47
23	1.20	0.74 - 1.95	1.20	0.73 - 1.97
24	0.78	0.48 - 1.26	0.78	0.48 - 1.29
25	0.33	0.20 - 0.54	0.32	0.19 - 0.55
26	0.59	0.36 - 0.96	0.58	0.35 - 0.95
27	0.88	0.54 - 1.43	0.89	0.54 - 1.45
28	0.53	0.32 - 0.87	0.54	0.33 - 0.90
29	1.14	0.70 - 1.86	1.13	0.69 - 1.85
30	0.49	0.30 - 0.80	0.48	0.29 - 0.79
31	0.53	0.32 - 0.86	0.51	0.31 - 0.84
32	0.93	0.57 - 1.50	0.94	0.57 - 1.54

values of α on different days, though β appears to be practically constant.

8. The Consequences of Pooling Data

Having established a need for the contemporary trial of S and T , one might still think that advantage must come from also pooling evidence on S from some previous assays. A look at a range of options in planning assays seems worth while. For this purpose it will suffice to work with "variances" of the m_i ; since it is estimated from a ratio, m_i does not have a true variance, but these assays are sufficiently precise to permit an asymptotic variance to be used satisfactorily. If g is negligibly small, equation (I.5.11) is equivalent to using (III.2.2); then

$$Var(m) = s^2(v_{11} - 2mv_{12} + m^2 v_{22})/b^2 , \qquad (IV.8.1)$$

and this can be adapted to quantal assays.

Suppose that an assay is to be essentially as described for pertussis vaccine, except that it will have n_1 mice at each of the three doses of S , n_2 mice at each of the three doses of T , and that β is regarded as constant. The proposal for examination is that the constant α_j for the response curve to the standard preparation on occasion j shall be estimated by pooling data for S with those from the previous (w - 1) assays. Current practice is to have $n_1 = n_2 = 32$, w = 1 ; interest is focused on decreasing n_1 , increasing w , and possibly increasing n_2 . Suppose also that the variance of an estimate of ImD50 can be regarded as made up of ϕ^2 from occasion to occasion plus an intra-assay component equivalent to σ^2 if there were only one mouse per dose. This cannot hold exactly for small n because the asymptotic variance properties would break down, but it is adequate for comparing, say, $n_S = 20$ and $n_S = 30$ in terms of variance components $\sigma^2/20$ and $\sigma^2/30$. Greater refinement here would mean an effectively greater decline in precision as n is reduced than the following argument implicitly assumes.

Essentially, we can estimate $\sigma^2/32$ from the maximum likelihood calculations by applying (IV.8.1) at the last stage of the maximization. Averaging over the 32 estimations of m_i used for Table 7.1 gave

$$\hat{\sigma}^2/32 = 0.01176$$

(individual values were all between 0.0114 and 0.0134). Moreover, a mean square deviation calculated for the 32 values of m_i estimates a total of inter and intra-day variance as

$$\hat{\phi}^2 + \hat{\sigma}^2/32 = 1.7790/31 = 0.05739 \ .$$

Hence

$$\left.\begin{aligned} \hat{\sigma}^2 &= 0.3763 \ , \\ \hat{\phi}^2 &= 0.04563 \ . \end{aligned}\right\} \qquad \text{(IV.8.2)}$$

For the proposed general design of assay

$$\text{Var}(m_S) = (\phi^2 + \frac{\sigma^2}{n_1})/w \quad . \qquad \text{(IV.8.3)}$$

Assuming the same intrinsic variability for T ,

$$\text{Var}(m_T) = (\phi^2 + \frac{\sigma^2}{n_2}) \quad . \qquad \text{(IV.8.4)}$$

Because of the contemporaneity of the T data with part of the S ,

$$\text{Cov}(m_S, m_T) = \phi^2/w \quad . \qquad \text{(IV.8.5)}$$

The use of a common β should in fact modify the covariance a little, but for good assays with precise estimation of β the effect is small.

Now the estimated log relative potency, equivalent to (I.3.9), is

$$M = m_S - m_T , \qquad (IV.8.6)$$

and

$$Var(M) = Var(m_S) - 2\,Cov(m_S, m_T) + Var(m_T)$$

$$= \phi^2(1 - \frac{1}{w}) + \sigma^2(\frac{1}{wn_1} + \frac{1}{n_2}) , \qquad (IV.8.7)$$

which of course simplifies to $\sigma^2/16$ for the existing practice of $n_1 = n_2 = 32$, $w = 1$. The ϕ^2 term, which disappears only if $w = 1$, is apt to dominate the behaviour of $Var(M)$ if $w > 1$ and n_1, n_2 exceed about 10.

For fixed n_1, n_2, as w increases,

$$Var(M) \to \phi^2 + \frac{\sigma^2}{n_2} ; \qquad (IV.8.8)$$

this limit is approached from above if

$$\phi^2 < \sigma^2/n_1 , \qquad (IV.8.9)$$

but otherwise from below.

Now (IV.8.9) is satisfied for the pertussis data (using $\hat{\sigma}^2$, $\hat{\phi}^2$) only if $n_1 \leq 8$, which would almost certainly be regarded as unreasonably small for n_1. Another form of argument is that a symmetric assay depending entirely upon contemporary comparisons will have $n_1 = n_2 = n$, $w = 1$, and therefore

$$Var(M) = 2\,\sigma^2/n .$$

On the other hand with small n_1 and large n_2, no amount of increase in the pooling of data (i.e. no increase in w) can bring $\mathrm{Var}(M)$ below ϕ^2. With the estimates in (IV.8.2), there can be some advantage for the latter proposal if $n \le 16$, but no appreciable advantage unless n were much smaller. If we further insist that $n_1 + n_2 \le 2n$, so that any improved precision of M is not obtained by expending more mice per occasion of assay, the minimal n for which the revised scheme has any advantages becomes even smaller. For example, $n = 10$, $n_1 = 2$, $n_2 = 18$ would represent equality of usage of mice. A reduction in $\mathrm{Var}(M)$ would then require

$$\phi^2 \left(1 - \frac{1}{w}\right) + \sigma^2\left(\frac{1}{2w} + \frac{1}{18}\right) < \frac{\sigma^2}{5} \quad ;$$

with the estimates available this is not satisfied until $w \ge 17$, and even a much larger w can at most reduce $\mathrm{Var}(M)$ from 0.075 to 0.066.

No more need be said on the pertussis evidence. There is no indication that any method of adjusting the numbers of subjects per preparation and pooling the information on S from previous assays can improve the precision of potency estimation (of course subject to using the same total number of mice). The examination of analogous data for another assay technique, whether with quantitative or with quantal responses, might tell a different story, but useful gains could be achieved only if ϕ^2/σ^2 is relatively small.

BIOASSAY

LECTURE V: ASSAYS OF ANTISERUM ACTIVITY

1. Antiserum Potency

"Immunological methods are more and more widely used for the assay of protein and polypeptide hormones. An important prerequisite for such assays is the use of specific antisera. Until now the specificity of antisera has been assessed only by immunological procedures and qualitative tests in animals. Were precise and statistically valid bioassay procedures available, it would be possible to provide an estimate of the biological potency of antisera and to express it in reproducible units. The development of such methods would also enable investigators to establish the quantitative profile of antibodies against different antigenic determinants carrying biological activities and to control in this way the biological specificity of various antisera". (Robyn, Diczfalusy and Finney, 1968).

Since the primary interest of an antiserum lies in its specific capacity for neutralizing the potency of a serum, a reasonable basis for measuring antiserum potency should be its neutralizing power. This implies a need to combine assay results in a manner quite different from that in Lecture III. Dose-response curves based on different amounts of antiserum can be compared, and from their combined evidence a numerical assessment of potency can be made.

2. The Additive Model

Robyn *et al.* developed a model that provides a first approximation to the need. Suppose that a hormone H can be neutralized by an antiserum AS, and that H is measured on an agreed scale (which may be in arbitrary "international units"). Define the antiunit (AU) of AS to be such that 1.0 AU neutralizes the specific biological activity of 1.0 units of H . Under a condition that we term *simple additivity,* a dose consisting of z_H units of H with z_S ml of AS will act exactly as $(z_H - \theta z_S)$ units of H alone, where θ is the neutralizing potency of the antiserum expressed in AU/ml . Hence a composite dose consisting of

$$\frac{z}{\gamma} \text{ units of H }, \frac{z(1-\gamma)}{\theta\gamma} \text{ ml of AS }, \qquad (V.2.1)$$

for any z and for any γ with $0 < \gamma \leq 1.0$, should act as z units of H alone. The composite dose is such that a fraction γ of H is unneutralized whatever the value of z .

The assay problem here is that of estimating θ . The additivity rule is in reality a gross oversimplification of the quantitative effect of antiserum in neutralizing a hormone, and biological theory indicates that it must break down in extreme circumstances. Nevertheless, at moderate levels of neutralization it appears to behave satisfactorily.

3. Basis for an Assay

By assaying a mixture of H and AS against H alone, θ can be estimated. If T_0 is a guessed value for θ , doses of a mixture corresponding to that defined in (V.2.1) can be taken as

$$\frac{z}{\gamma} \text{ units of H }, \frac{z(1-\gamma)}{T_0\gamma} \text{ ml of AS }, \qquad (V.3.1)$$

with an agreed value of γ and a sequence of levels of z . The hormone alone of course corresponds to taking $\gamma = 1$. As a simple assay design, one might take three doses of H alone in the ratio 1:2:4 and three doses of the mixture defined by (V.3.1) with values of z in the same ratio; any pre-existing knowledge of potency should be used to make corresponding doses produce nearly equal responses. If additivity obtains, the true potency of the mixture, relative to H, is

$$\rho = \frac{1}{\gamma} - \frac{\theta(1-\gamma)}{T_0\gamma}$$

$$= 1 + \left(1 - \frac{\theta}{T_0}\right)\frac{1-\gamma}{\gamma} . \qquad (V.3.2)$$

If R is the estimate of ρ, obtained by standard statistical analysis, equation (V.3.2) can be inverted to give T, an estimate of θ, as

$$T = T_0(1 - R\gamma)/(1 - \gamma) , \qquad (V.3.3)$$

which can be written

$$T = T_0 + \frac{T_0\gamma(1-R)}{1-\gamma} \qquad (V.3.4)$$

to display the departure from T_0. Naturally $R = 1$ gives $T = T_0$. Moreover, if R_L, R_U are limits to ρ, calculated by Fieller's theorem at a selected probability, they can be used to give limits to T:

$$\left.\begin{aligned} T_L &= T_0(1 - R_U\gamma)/(1 - \gamma) , \\ T_U &= T_0(1 - R_L\gamma)/(1 - \gamma) . \end{aligned}\right\} \qquad (V.3.5)$$

The factor $(1 - R\gamma)$ is the fraction of the hormone content estimated to be neutralized in each composite dose.

The relation between R_L, R_U and T_L, T_U is such that the range associated with θ may be entirely different from that for ρ. When γ is small, and neutralization therefore almost complete, the value of R has little effect on T and even a wide difference between R_L and R_U will leave T with high precision. This indication of the desirability of using a small γ must not be accepted too readily; if neutralization is to be nearly complete, technical errors in the measurement of doses and in the preparation of dilutions become critically important. In practice, γ of the order of 0.5 seems satisfactory, but the statistical attractions of $\gamma < 0.1$ must be balanced against both the risk of technical errors and the possible breakdown of additivity. If γ is large, the precision of T will be adversely affected by even small errors of estimation of R.

Robyn and Diczfalusy (1968a) described an assay of an anti-gonadotrophic serum. They had doses of 1.0, 2.0, 4.0 international units of human chorionic gonadotrophin (HCG), and also composite doses consisting of 2.0 units HCG + 1/1280 ml AS, 4.0 units + 1/640 ml, and 8.0 units + 1/320 ml ; the composite doses were thought to correspond to 50% neutralization. A randomized block design used 5 litters of 6 immature male rats; the response measured was the total weight of the accessory reproductive organs. The assay led to $R = 0.772$, with 0.95 limits at 0.647, 0.911. Since the composite doses had been based on $T_0 = 1280$ and $\gamma = 0.5$, equation (V.3.3) gives

$$\begin{aligned} T &= 1280(1 - 0.772 \times 0.5)/0.5 \\ &= 1570 . \end{aligned}$$

Thus the assay leads to the estimate that the antiserum contained 1570 AU/ml, with limits (probability 0.95) similarly calculated as 1390 AU/ml and 1730 AU/ml .

Robyn and Diczfalusy (1968a, b) collated, from many similar assays, estimations of the neutralizing potency of various antigonadotrophic sera against HCG and also against luteinizing hormone (LH) and follicle stimulating hormone (FSH). They discussed the practical limitations of these techniques.

4. Assay Validity

In §I.4, I emphasized the need to incorporate tests of validity of the underlying model into the design and analysis of bioassays. The condition of simple additivity clearly is of crucial importance to the estimation, and no assay system can be considered adequate if it makes no provision for testing validity. A single assay of the kind I have described cannot do this, except in so far as a gross departure from additivity might cause the regression for the composite doses to deviate from parallelism with that for the hormone alone or to become non-linear.

The solution appears to lie in testing composite doses based on two or more values of γ. If T is calculated from each series of composite doses based on a common γ, a plot of T against γ should either reveal any systematic trend that arises from a breakdown of additivity or confirm the hope that T is practically constant. Converting the visual examination into a formal test is more troublesome. Account must be taken of whether a multiple assay involved measurement of responses for several values of γ within one experiment, or whether each of a series of separate experiments relates a sequence of doses based on one value of γ to doses of H alone.

Suppose that a multiple assay has been constructed, with H alone as the standard preparation, and k test preparations are represented by series of composite doses based on γ_i $(i = 1,2, \ldots, k)$. A consequence of additivity and the standard theory of parallel line assays (equation I.3.5) is that the $(k + 1)$ regression lines shall be of the form (for $i = 0, 1, \ldots, k$)

$$Y = \alpha + \beta\log\{1 + (1 - \gamma_i)(T_0 - \theta)/T_0\gamma_i\} + \beta x \ , \quad \text{(V.4.1)}$$

where $\gamma_0 = 1$ for H alone. Since T_0 is fixed and the γ_i are chosen by the investigator, this implies adequacy of the three parameters (α,β,θ) to describe all the data (apart from a variance). If a separate θ_i were allowed for each γ_i , so that the basic analysis can be in terms of estimates of relative potency for each preparation, a total of $(k + 2)$ parameters would be needed.

Estimation of the $(k + 2)$ parameters presents no difficulty. In the ordinary manner the responses are fitted with $(k + 1)$ parallel lines of the form

$$\left.\begin{aligned} Y_S &= \hat{\alpha}_S + \hat{\beta}x \\ Y_i &= \hat{\alpha}_i + \hat{\beta}x \quad (i = 1,2, \ldots, k) \ ; \end{aligned}\right\} \quad \text{(V.4.2)}$$

then for the "pseudo test preparation" based on γ_i , an estimator, T_i, of the neutralizing potency θ_i is obtained as the unique solution of

$$\hat{\beta}\log\{1 + (1 - \gamma_i)(T_0 - T_i)/T_0\gamma_i\} = \hat{\alpha}_i - \hat{\alpha}_S \ . \quad \text{(V.4.3)}$$

The T_i are not independent, but inspection of a plotting of T_i against γ_i will at least give a clue to whether or not θ_i is constant. The restricted estimation of 3 parameters involves minimizing a sum of squares for a non-linear model essentially as in §II.5. The difference between the two residual sums of squares is a sum of squares with $(k - 1)$ degrees of freedom appropriate to testing the hypothesis that the θ_i

are equal, that is to say testing additivity. I see no merit in any large value of k, and probably $k = 4$ is adequate for most purposes. Robyn (1969) and Petrusz *et al.* (1970) presented empirical evidence that mixtures of HCG with anti-HCG and of FSH with anti-FSH conform well to predictions based on additivity. They used experiments with γ between 0.8 and 0.01 but found anomalous behaviour for $\gamma \leq 0.1$ either because of technical flaws or because of new factors modifying the theory.

The method of estimation and testing requires minor modifications if k separate assays, one for each γ_i, replace the one multiple assay. In this situation, an approximation should suffice if all assays are very precise, so that g (as defined in §I.5) is always small and a variance of log potency can be used (cf. III.2.2). From the usual 'differential' approximations to the variances of functions, we can obtain successively for assay i

$$\mathrm{Var}(R_i) = R_i^2 \, \mathrm{Var}(M_i)$$

and, from equation (V.3.3),

$$\mathrm{Var}(T_i) = T_0^2 \, \gamma_i^2 \, \mathrm{Var}(R_i)/(1 - \gamma_i)^2 \, . \qquad \text{(V.4.4)}$$

The constancy of θ can then be tested by the approximate

$$\chi^2_{[k-1]} = \sum_i W_i T_i^2 - \left(\sum_i W_i T_i\right)^2 / \sum_i W_i \qquad \text{(V.4.5)}$$

where

$$W_i = 1/\mathrm{Var}(T_i) \, . \qquad \text{(V.4.6)}$$

If no serious disagreement is disclosed,

$$\bar{T} = \sum_i W_i T_i / \sum_i W_i \qquad \text{(V.4.7)}$$

is a weighted mean of the T_i's and

$$\mathrm{Var}(\bar{T}) = 1/\sum_i W_i . \qquad \text{(V.4.8)}$$

The approximations inherent in this procedure seem likely to exaggerate the precision at all stages, and hence to inflate χ^2 - so tending to indicate heterogeneity of estimates where in reality additivity obtains. Romani *et al.* (1974) used (V.4.4) - (V.4.8) in many comparisons of the potencies of anti-HCG sera in neutralization of the hormone with $\gamma = 0.5$ and $\gamma = 0.1$, and found few instances of appreciable heterogeneity.

As in Lecture III, this weighted mean procedure can be improved by following the maximum likelihood approach. The algebra will be less tractable than for the earlier problem of combining potency estimates, but no numerical difficulty should be encountered. One may speculate that again multiple maxima could occur.

5. Neutralization of Antiserum

An essential feature of the hormone-antiserum situation, not explicitly mentioned earlier, is that the only property of the antiserum concerning the investigator is its power to meutralize the hormone. Thus the additive rule operates only when $(z_H - \theta z_S)$ is non-negative; an excess of antiserum merely leaves a state of complete neutralization. Of course, $\gamma \leq 1.0$ is also a necessary condition. The research in association with which the statistical ideas described earlier were devised disclosed that certain gonadotrophin preparations have an immunological capability but show little or no biological activity. Such a preparation can neutralize antiserum, but cannot contribute to the biological activity of the hormone itself. The potency of this *antiserum neutralizing factor* (ANF) can be measured in anti-antiunits (AAU), where 1 AAU combines with and neutralizes the effect of 1 AU of antiserum. The ideas of §V.2 must be extended . A composite dose consisting of, say, z_H units of H , z_S ml of AS, and z_N mg of ANF will act as though it were a dose

$$(z_H - \theta_S z_S + \theta_S \theta_N z_N) \text{ units of } H . \qquad (V.5.1)$$

This is subject to two limitations arising from the inactivity of excess amounts of AS and ANF; if $(z_S - \theta_N z_N) < 0$, the composite dose acts as though it were simply z_H , and if the residual antiserum potency after neutralization by ANF is more than enough to neutralize all the hormone then the composite dose is inactive. In other words, the formula in (V.5.1) is to be interpreted as z_H or 0 if its magnitude exceeds z_H or is less than zero respectively.

By generalization of (V.2.1), three equivalent doses can be defined:

units of H	ml of AS	mg of ANF
z	0	0
$\frac{z}{\gamma_S}$	$\frac{z(1-\gamma_S)}{\theta_S \gamma_S}$	0
$\frac{z}{\gamma_S}$	$\frac{z(1-\gamma_S)}{\theta_S \gamma_S \gamma_N}$	$\frac{z(1-\gamma_S)(1-\gamma_N)}{\theta_S \theta_N \gamma_S \gamma_N}$

with the requirement that $0 < \gamma_S \leq 1.0$, $0 < \gamma_N \leq 1.0$ where γ_S, γ_N are the two fractions unneutralized. Here θ_S is the potency of AS, previously denoted by θ , and $\theta_S \theta_N$ is the potency of ANF in AAU per mg. In order to estimate $\theta_S \theta_N$, in the usual situation that θ_S is not known and must be simultaneously estimated, the minimal experiment is one that tests H alone and the two types of composite dose at various levels of z . Provisional guesses of T_{S_0} , T_{N_0} might replace θ_S, θ_N in the formulae, and values of γ_S , γ_N might be taken as about 0.5 each. Alternatively, two assays might be used, one comparing H alone with a sequence of composite doses of H and AS (i.e. γ_S fixed and $\gamma_N = 1.0$) and the second comparing this type of composite with the triple composite having γ_S and γ_N fixed but both less than 1.0 .

Let R_S denote the potency estimated for the first composite relative to H and R_N that of the second composite relative to the first. For the first of the two assays, the equations of §V.3 still hold, with insertion of the subscript S throughout. The second potency estimate leads to the equation

$$R_N = [1 - \frac{T_S(1-\gamma_S)}{T_{S_0}\gamma_N} + \frac{T_S T_N(1-\gamma_S)(1-\gamma_N)}{T_{S_0}T_{N_0}\gamma_N}] \;/\; [1 - \frac{T_S(1-\gamma_S)}{T_{S_0}}] \;, \qquad (V.5.2)$$

easily obtained but scarcely very tractable. Combination of the two estimating equations, followed by some rearrangement of terms, leads to

$$T_S T_N = T_{S_0} T_{N_0} [1 + \frac{\gamma_S(1-R_S) - \gamma_S\gamma_N(1-R_S R_N)}{(1-\gamma_S)(1-\gamma_N)}] \qquad (V.5.3)$$

as an analogue of (V.3.4) .

For an equation as complicated as (V.5.3), any exact theory of probability limits, either for θ_N alone or for the product $\theta_S\theta_N$, is most unlikely to be obtainable. The log likelihood of the data can readily be written as a function of θ_S and θ_N . Its maximum will occur for T_S, T_N , since these are uniquely related to R_S,R_N, and intervals for θ_S, θ_n jointly could be obtained, at least asymptotically, by regarding twice the difference of the log likelihood from its maximum as a $\chi^2_{[2]}$. If doses have been well chosen and dose-response regression coefficients are large enough to ensure high precision, the simpler procedure of calculating approximate variances based upon differentials should be adequate. With R_S, R_N obtained from distinct experiments, this leads to

$$\mathrm{Var}(T_S T_N) = \left(\frac{T_{S_0} T_{N_0} \gamma_S}{(1-\gamma_S)(1-\gamma_N)} \right)^2 [(1 - \gamma_N R_N)^2 \mathrm{Var}(R_S) + \gamma_N^2 R_S^2 \mathrm{Var}(R_N)] \ . \tag{V.5.4}$$

Alternative experimental procedures can be considered. Instead of being as described above, the second assay might compare H alone with the triple composite involving H, AS and ANF, so giving a direct estimate of the product $R_S R_N$. Equation (V.5.3) would still be needed for the ANF potency, but equation (V.5.4) would be replaced by the somewhat simpler form

$$\mathrm{Var}(T_S T_N) = \left(\frac{T_{S_0} T_{N_0} \gamma_S}{(1-\gamma_S)(1-\gamma_N)} \right) [\mathrm{Var}(R_S) + \gamma_N^2 \mathrm{Var}(R_S R_N)] \ . \tag{V.5.5}$$

On the other hand, a single experiment on all three dose series would give greater complexity on account of covariance terms. Not even an approximate theory of design for optimal precision has been studied.

Once again validity testing is important, now in respect of the additivity of ANF. Equation (V.4.1) generalizes in obvious fashion, and experimentation with dose series based on different values of γ_N can be used in association with additional non-linear least squares calculations to test the assumption that additivity obtains. Here no new difficulties enter.

Petrusz *et al.* (1971a, b) first described this theory, using slightly different notation, and illustrated the calculations from studies on human chorionic and menopausal gonadotrophins. They found full least squares analysis to be scarcely necessary, as cruder methods sufficed to indicate the important features of the results, including failure of the additivity condition in some circumstances. Qazi *et al.* (1974) produced extensive evidence that the phenomena associated with the neutralization of antisera, the nature of which some have challenged, do in fact occur.

The mathematician may like to think of the next stage, the antiserum neutralization neutralizing factor, with a potency measured in anti-anti-antiunits. I do not think biologists will seek evidence for it enthusiastically; in the unlikely event of its existence, estimation of its potency may prove inherently very imprecise.

REFERENCES

Armitage, P. (1970) The combination of assay results. *Biometrika*, 57, 665-666.

Armitage, P. and Bennett, B.M. (1974) Maximum likelihood solutions for the combination of relative potencies. *Journal of Hygiene*, 73, 97-99.

Armitage, P., Bennett, B.M., and Finney, D.J. (1976) Point and interval estimation in the combination of bioassay results. *Journal of Hygiene*, 76, 147-162.

Barnett, V.D. (1966) Evaluation of maximum likelihood estimator where the maximum likelihood equation has multiple roots. *Biometrika*, 53, 151-165.

Bennett, B.M. (1963) On combining estimates of a ratio of means. *Journal of the Royal Statistical Society*, B25, 201-205.

Brenner, P.F., Guerrero, R., Cekan, Z., and Diczfalusy, E. (1973) Radioimmuno methods for six steroids in human plasma. *Steroids*, 21, 775-794.

Edwards, A.W.F. (1972) *Likelihood*, Cambridge: The University Press.

Ekins, R. and Newman, B. (1970) Theoretical aspects of saturation analysis. *Acta Endocrinologica*, *suppl.* 147, 11-36.

Fieller, E.C. (1944) A fundamental formula in the statistics of biological assay, and some applications. *Quarterly Journal of Pharmacy and Pharmacology*, 17, 117-123.

Finney, D.J. (1964) *Statistical Method in Biological Assay*. (2nd edition). London: Charles Griffin and Co. Ltd.

Finney, D.J. (1971) *Probit Analysis*. (3rd edition). Cambridge: The University Press.

Finney, D.J., Holt, L.B. and Sheffield, F. (1975) Repeated estimations of an immunological response curve. *Journal of Biological Standardization*, 3, 1-10.

Genesis, Chapter 8, Verses vi-xi.

Healy, M.J.R. (1972) Statistical analysis of radioimmunoassay data. *Biochemical Journal*, 130, 207-210.

Kendrick, P.L., Eldering, G., Dixon, M.K. and Misner, J. (1947) Mouse protection tests in the study of pertussis vaccine. *American Journal of Public Health*, 37, 803-812.

McArthur, J.W., Ulfelder, H. and Finney, D.J. (1966) A flexible computer program for the composite analysis of symmetrical and asymmetrical biologic assays of parallel-line type. *Journal of Pharmacology and Experimental Therapeutics*, 153, 573-580.

Midgley, A.R., Niswender, G.D. and Rebar, R.W. (1969) Principles for the assessment of the reliability of radioimmunoassay methods (precision, accuracy, sensitivity, specificity). *Acta Endocrinologica, suppl.* 142, 163-184.

Nelder, J.A. and Mead, R. (1965) A simplex method for function minimization. *Computer Journal*, 7, 308-313.

Petrusz, P., Diczfalusy, E. and Finney, D.J. (1971a) Bioimmunoassay of gonadotrophins. 1. Theoretical considerations. *Acta Endocrinologica*, 67, 40-46.

Petrusz, P., Diczfalusy, E. and Finney, D.J. (1971b) Bioimmunoassay of gonadotrophins. 2. Practical aspects and tests of additivity. *Acta Endocrinologica*, 67, 47-62.

Petrusz, P., Robyn, C., Diczfalusy, E. and Finney, D.J. (1970) Bioassay of antigonadotrophic sera. 4. Experimental verification of the principle of additivity. *Acta Endocrinologica*, 63, 150-160.

Qazi, M.H., Romani, P. and Diczfalusy, E. (1974) Discrepancies in plasma LH activities as measured by radioimmunoassay and an "in vitro" assay. *Acta Endocrinologica*, 77, 672-686.

Robyn, C. (1969) Biological and immunological characterization of anti-gonadotrophic profiles. *Acta Endocrinologica, suppl.* 142, 31-53.

Robyn, C. and Diczfalusy, E. (1968a) Bioassay of antigonadotrophic sera. 2. Assay of the human chorionic gonadotrophin (HCG) and luteinizing hormone (LH) neutralizing potencies. *Acta Endocrinologica*, 59, 261-276.

Robyn, C. and Diczfalusy, E. (1968b) Bioassay of antigonadotrophic sera. 3. Assay of the human follicle stimulating hormone (FSH) neutralising potency. *Acta Endocrinologica*, 59, 277-297.

Robyn, C., Diczfalusy, E. and Finney, D.J. (1968) Bioassay of antigonado-trophic sera. 1. Statistical considerations and general principles. *Acta Endocrinologica*, 58, 593-599.

Rodbard, D. (1971) Statistical aspects of radioimmunoassays. In *Principles of Competitive Protein Binding Assays*. (Philadelphia: Lippincott), 204-259.

Rodbard, D. and Cooper, J.A. (1970) A model for prediction of confidence limits in radioimmunoassays and competitive protein binding assays. *Proceedings of Symposium on Radioisotopes in Medicine*, Vienna 1970, 659-674.

Rodbard, D. and Frazier, G.R. (1974) Statistical analysis of radioligand assay data. In *Methods in Enzymology* (New York: Academic Press)

Rodbard, D. and Hutt, D.M. (1974) Statistical analysis of radioimmunoassays and immunoradiometric (labeled antibody) assays: a generalized weighted, iterative, least-squares method for logistic curve fitting. *Symposium on RIA and Related Procedures in Medicine*, 1, 165-192.

Romani, P., Robyn, C., Petrusz, P. and Diczfalusy, E. (1974) Bioassay of antigonadotrophic sera. 5. Further studies on the reliability of the bioassay method for the estimation of human chorionic gonadotrophin (HCG) neutralizing potency. *Acta Endocrinologica*, 76, 629-644.

Sprent, P. (1966) A generalized least-squares approach to linear functional relationships. *Journal of the Royal Statistical Society*, B28, 278-297.

Qualitative Analysis of Complex Systems

Professor R. Levins

Centre for Applied Science
School of Public Health
Harvard University
Boston, Massachusetts

QUALITATIVE ANALYSIS OF COMPLEX SYSTEMS

I

At present ecology is caught between strong opposing forces which both stimulate and inhibit the development of an effective, complex approach to the basic theoretical issues and the problems of practice.

The internal development of the biological disciplines increasingly demands integration. It is apparent that the time scales of biogeographic, demographic, and evolutionary events may often overlap so that the competition or predator-prey relations among species are imbedded in cycles of local extinctions and recolonization. It is also apparent populations adapt while they interact. Therefore we must move toward a theory in which populations are heterogeneous for genotype, age structure, and physiological condition. To treat the environment as given (and represented by a selection coefficient or demographic parameter) without considering how organisms select, define, modify and respond to complex patterns of environment heterogeneity is likewise proving unsatisfactory. In principle therefore, ecologists recognize their discipline as one of complexity and interaction whether or not they deal with this complexity in their research.

The problems which society has placed before ecology and which we have been least successful in confronting are themselves problems of complex interactions. World food, resource management, environmental protection, and public health problems are all obviously interdisciplinary and cry for broad holistic approaches. However two sets of factors operate against such a program. First, there is the social organization of scientific work. Increasingly, scientific research, even in ecology, has become a commodity, produced for sale. The research carried on in the laboratories of the large corporations is clearly determined by considerations of profitability, both in the allocation of investment among possible research projects and in the allocation between R & D expenditures as a whole and alternatives such as greater sales effort, mergers, or bribery. In the context of a cost-accounting evaluation, the scope of the research effort is limited to what is economically relevant, and the division of labour among laboratories makes it unlikely that scientists will intervene directly in examining the interactions between their work and that of other fields.

Thus the boundary conditions of a problem are simply handed to the researcher and accepted as given. But to accept boundary conditions is often to reinforce them. Agricultural engineers design machines for work in large scale monoculture because that is how potential customers farm. Plant breeders select for performance in pure stands because that is what the machinery can handle. Pure stands are recommended because that is where the varieties do best. Or, automobiles make shopping plazas possible and then shopping plazas make autos necessary. This kind of positive feedback (or vicious circle) is not the result of any conscious decision by researchers. Rather, it is a consequence of the contradiction between the growing sophistication at the smallest level, within the laboratory, and the greater irrationality in the broad perspective.

The division of labour in an enterprise or agency exacerbates these trends. The same division is converted into a philosophy of nature modeled after the organization of one's own place of work. The world becomes a list of discrete problems, to each of which there corresponds a magic bullet. Because investigators understand small areas with increasing detail, and feel less competent outside those areas, they reinforce the fetish of precision and become increasingly uncomfortable with qualitative ideas or variables that are difficult to measure. Therefore, their own philosophy lends force to the impositions of the organization of scientific labour, while many variables, such as the anxiety of a patient whose blood sugar is easily measured or the recreational importance of a forest whose cubic feet of timber is calculable, are relegated to footnotes.

Further, the noxious division of labour into applied and pure research encourages a peculiar mixture of short-sighted empiricism and irrelevant abstraction.

Finally, two other factors enter to facilitate whole system, complex approaches to phenomena. First, the mathematical sophistication of biologists has grown so that the level of complexity which is manageable has increased. Secondly, biologists are showing an increasing awareness of and interest in the philosophical issues in their sciences. One consequence of the U.S. political upheavals of the 1960's has been that U.S. scientists are now at least a little more familiar with the marxist tradition which, developing largely outside of academic science, has always stressed complexity, interaction, heterogeneity, and process.

We are left, then, with a fluid situation in which strong pressures are at work both demanding and preventing a holistic approach to the issues of ecology in the broad sense. These lectures are intended to demonstrate the possibility of facing complex systems, of working with variables that are not readily measured, and of reaching conclusions concerning the direction of efforts needed.

If the demand for a more complex view of the systems we study only meant the addition of more variables or parameters, matters would very soon become unmanageable. Therefore, the admission of additional components into our systems must be accompanied by the opposite process of reducing the number of variables needed. This can be approached in the following ways:

1. The search for macroscopic, composite measures in ecological systems which can replace the study of the original observables.
2. The stratification of a system into partly autonomous subsystems on the basis of rates of change rather than the physical nature of the variables. Then, from the point of view of the analysis of one set of variables, those which are much slower are treated provisionally as constants while those which are faster are assumed to have already reached their asymptotic behavior, which is some function of the states of the variables of interest.
3. The development of algorithms for the lumping of variables.
4. The development of methods which estimate the behaviour of the system using general structural properties such as connectivity or statistical measures of the interaction patterns.

In what follows we will touch on each of these approaches. First, however, it is necessary to comment on the use of linear tools and the mathematics of equilibrium properties in a world which we know is not linear and not at equilibrium.

The observation of regularities in nature implies that some variables are near equilibrium with respect to others even as the equilibrium points themselves move. Thus, a population may approach a stable age distribution even while it grows; the genotype frequencies of a population at the time of reproduction may be in Hardy-Weinberg equilibrium while natural selection changes gene frequencies; orthophosphate may be in equilibrium with other chemical forms while algae deplete the phosphate. This suggests that many changes can be understood in terms of the shifting of relative, subordinate equilibria.

Second, and more generally, any trajectory can be described by constants of motion which are more or less constant compared to the motion of the original variables. Therefore we can study these constants of motion as descriptors of systems.

In fact, motion and equilibrium are not the mutually exclusive opposites implied in Zeno's paradox. Rather, equilibrium is a form of motion, and it is important to find out how useful this understanding will be in the study of motion.

Macroscopic variables which describe whole-system properties arise in several ways. Some emerge prior to an investigation as objects of practical or theoretical interest such as diversity, biomass, or net productivity. Initial notions, which are often vague, must first be translated into a measureable quantity which satisfies heuristic conceptions regarding its behavior. Sometimes several alternative measures are proposed for the same property of systems. For instance, diversity may be measured by the number of species in a community, by the information measure

$$H = -\sum p_i \log p_i \tag{2.1}$$

where p_i is the frequency of species i in the community, by the antilog e^H, or by the measure

$$D = 1/\sum {p_i}^2. \tag{2.2}$$

The first measure, the number of species, ignores relative abundances. But for that reason it can be computed from the records of collecting expeditions of the past which gathered systematic and biogeographic rather than ecological information. The other measures, (2.1) and (2.2), consider both the number of species and their equability. They are maximal when the species are equally abundant and decrease as the distributions depart from uniformity. Field experience indicates that these measures can be evaluated quite rapidly, even when a majority of the species remains uncollected and they are relatively insensitive to taxonomic errors involving rare species. The measure H has the special virtue that it readily permits a components of diversity analysis in which total diversity, H_T, can be decomposed as

$$H_T = \bar{H}_W + H_A \tag{2.3}$$

where $\bar{H}_W$ is the average diversity within collections and H_A is the diversity due to differences among collections. However, the form of the H measure, which is the same as that for entropy in physics and information in communications engineering, inevitably hides the meaning in a cloud of metaphor; H is not the information or complexity or organization of a community. It is simply one of several measures of diversity.

The D measure is readily calculated in the field (this advantage diminishes as battery-operated calculators become available), is free of the many meanings associated with H, and constitutes part of a system of measures based on first- and second-order statistics that includes niche breadth and species overlap.

But questions of convenience are less important than those of biological significance. Once diversity has been defined as a descriptor, we can look for regular patterns in its distribution. Is diversity relatively constant over time when individual species in a community fluctuate? Does it vary between the temperate and tropical regions, climax and secondary communities, islands and continents? Do herbivore and carnivore communities differ with respect to diversity in regular ways? Does it change during eutrophication? Does it correlate with habitat diversity or with other measures of the community such as average overlap?

If the measurement of diversity does in fact reveal regularities of any interesting kind, then diversity becomes an object of study in its own right. As an object of study it should be computed regularly, and the determinants of diversity should be explored.

A second kind of problem arises in considering the measurement of competition, or more generally of species interaction. Many measures can be used to satisfy our heuristic sense of overlap, and often these can then be plotted to answer directly questions about trends in species packing. But in order for a measure to be useful in the study of coexistence or stability it must be further restricted. Gause's α, which is used to measure competition in a two-species system, was defined as the relative reduction in the growth rate of one species due to an individual of the other species. Thus, if the variables x_i are species abundances and

$$dx_i/dt = f_i(x_1, x_2, \ldots) \tag{2.4}$$

we can define the competition coefficient of species j on species i as

$$a_{ij} = \{(\partial/\partial x_j)(dx_i/dt)\}/\{(\partial/\partial x_i)(dx_i/dt)\} . \tag{2.5}$$

This is consistent with Gause's usage. When evaluated at equilibrium, a_{ij} shows the effect of one species on the abundance of another; it also determines whether or not the equilibrium is stable in the two-species case.

But for three or more species the matrix of a_{ij}'s is not sufficient to determine stability. In this circumstance we must use the matrix of first partial derivatives of (2.4),

$$(\partial/\partial x_j)(dx_i/dt). \tag{2.6}$$

The original overlap measure has become two measures that are needed for different purposes. Fortunately, the matrix specified in (2.5) is related to that specified in (2.6) by row multiplication so that it is possible to study the properties of one through the other. But of course the definition of a_{ij} is not yet its measure.

Another class of macroscopic measures arises because they are needed to interpret earlier properties of interest. For instance, the local dynamics of a system depend on the eigenvalues of the matrix of first partial derivatives around that point. Therefore, the eigenvalues become objects of interest. They can, of course, be calculated directly if the coefficients of the matrix are known, or they can be inferred indirectly from incomplete information. The attempt to determine the eigenvalues indirectly prompts the definition of new entities called feedback measures; these lead, in turn, to measures of the connectivity of the system, and so on.

Consider first a community of n species whose equilibrium levels are given by the familiar equation

$$K_i = x_i + \sum_{j \neq i} a_{ij} x_j . \qquad (i = 1, \ldots, n)$$

Diversity is maximized if $K_i = K$ and $a_{ij} = a$ for all i,j; then

$$x_i = \frac{K}{1 + (n-1)a}$$

for $i = 1, \ldots, n$. Now perturb K_i and consider how uneven the distribution becomes. First,

$$\frac{\partial x_i}{\partial K_i} = \frac{D_{n-1}}{D_n},$$

where D_{n-1} and D_n are determinants with principal diagonal elements equal to unity and off-diagonal elements a. Since

$$D_n = (1-a)^{n-1}\{1+(n-1)a\}, \tag{2.7}$$

it follows that

$$\frac{\partial x_i}{\partial K_i} = \frac{1+(n-2)a}{(1-a)[1+(n-1)a]}.$$

Finally, the proportional change in x_i due to a change in K_i is

$$\frac{K_i}{x_i}\frac{\partial x_i}{\partial K_i} = \frac{1+(n-2)a}{1-a}.$$

Now consider $\frac{\partial x_j}{\partial K_i}$, the effect on other species of a change in K_i; clearly

$$\frac{\partial x_j}{\partial K_i} = \frac{A_{ij}}{D_n}$$

where A_{ij} is the cofactor of an off-diagonal element. But we are perturbing around a matrix in which all the cofactors of off-diagonal elements are equal to A, say. The value of A can be determined by expanding the determinant by the elements of a row or column; thus

$$D_n = D_{n-1} + (n-1)aA. \tag{2.8}$$

Using (2.7) in (2.8) we have

$$A = (1-a)^{n-2}[(1-a)\{1+(n-1)a\} - \{1+(n-2)a\}]/(n-1)a,$$

or finally

$$A = -(1-a)^{n-2}a.$$

Hence

$$\frac{K_i}{x_j}\frac{\partial x_j}{\partial K_i} = -\frac{a}{1-a}\;.$$

Thus, for a given, positive perturbation of K_i due to an arbitrary difference in the biology of species i, x_i increases and x_j decreases by amounts that increase with a. The more overlap, a, that exists, the more irregular will be the distribution of species and hence the smaller will be the diversity.

In the rest of this paper we will be investigating the uses of signed digraphs in the study of systems about which we have only partial information. The method is based on the researches of Mason (1953). See Roberts (1976) for a complete elementary exposition of the technique.

We begin by representing the dynamics of a community of n species by a set of differential equations

$$\frac{dX_i}{dt} = f_i(X_1, X_2, \ldots, X_n;\ C_1, C_2, \ldots), \quad (i = 1, \ldots, n) \tag{3.1}$$

the X_i's are species abundances and the C_j's are parameters of the system that depend on the environment and the biology of the component species. There may be one or more equilibrium points of the system at which all the derivatives dX_i/dt are zero. In the neighborhood of any equilibrium point the behavior of the system depends on the matrix

$$A = (a_{ij}) \tag{3.2}$$

where a_{ij} is equal to $\frac{\partial f_i}{\partial X_j}$ evaluated at the equilibrium point.

A graph of the system is formed by taking the variables X_i as vertices, and drawing oriented lines from X_j to X_i, each with the magnitude a_{ij}. Corresponding to diagonal elements a_{ii} there are loops of length 1 from X_i to itself.

The values of the a_{ij}'s may be difficult to obtain, and may be different at different equilibrium points; it is usually easy, however, to determine the sign of a_{ij}. Thus, if X_j preys on X_i then a_{ij} is negative and a_{ji} is positive, and we distinguish between positive and negative links by an arrow and a circle respectively.

Several kinds of ambiguities may arise. First, a pair of species may interact in more than one way. For instance, green algae and blue-green algae may compete for phosphate, blue-greens may fix nitrogen which then becomes available to the greens, green algae may secrete a vitamin that stimulates the blue-greens, blue-greens may produce a toxin that inhibits the greens and at high population densities they may deprive each other of light. Species may compete with each other for extraneous food but may also prey on each other's young. The categories of competition, predator-prey, symbiosis or commensalism are the simplified summaries of complex relations. When such ambiguities arise they can be removed by the addition of new variables. Thus, the system with blue-green and green algae must be extended to include at least nitrogen and phosphorus. In Figure 1 we show the graphs for several systems in which ambiguities have been removed by expanding the system.

A second ambiguity arises if the sign of the effect of a variable depends on its level, or the level of other variables. In these circumstances we must specify the range of values within which a given graph is thought to apply.

A third ambiguity relates to the self-damping terms a_{ii}, corresponding to loops of length 1. Suppose that the differential equation is of the form

$$\frac{dX_i}{dt} = X_i G_i(X_{j \neq i};\ C_1, C_2, \ldots), \quad (i = 1, \ldots, n)$$

in which X_i does not appear in G_i. Then

$$\frac{\partial}{\partial X_i}\left(\frac{dX_i}{dt}\right) = X_i \frac{\partial G_i}{\partial X_i} + G_i \ .$$

But $\frac{\partial G_i}{\partial X_i} = 0$ and at equilibrium $G_i = 0$ if X_i remains in the system.

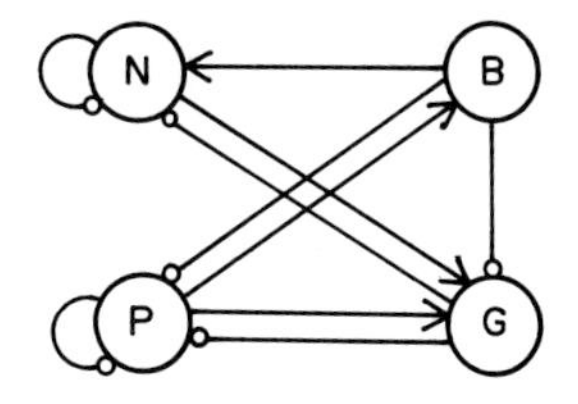

(a) The variables are nitrate (N), phosphate (P), blue-green (B) and green (G) algae. The graph includes competition by the algae for phosphate, inhibition of greens by blue-greens, consumption of nitrate by greens, and nitrogen fixation by blue-greens.

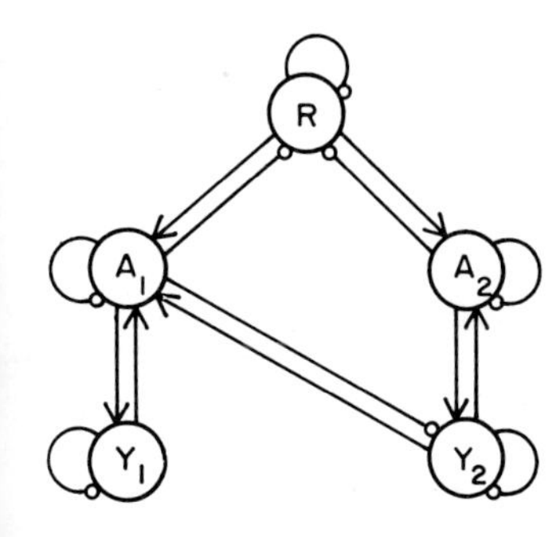

(b) Adults, (A_1 and A_2) of two species produce and are produced by their young (Y_1 and Y_2). The adults compete for resource R; in addition species 1 adults prey on the young of species 2.

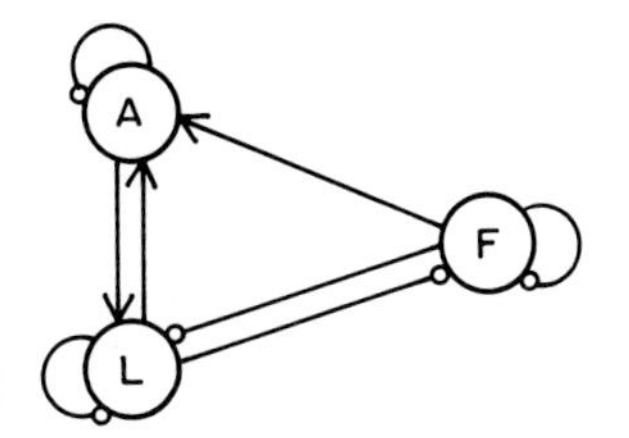

(c) In this representation of tree-hole mosquitoes, larval development (L) is speeded up by available food (F) in the water. Therefore food reduces the number of larvae and increases the number of adults (A).

Figure 1

Therefore, $a_{ii} = 0$, and a species with simple reproduction has no self-damping loop.

Note, however, that if a species' rate of reproduction decreases because of crowding then a_{ii} will be negative, while if the rate is autocatalytic a_{ii} may be positive. The autocatalytic situation may arise when the population density is very low and the probability of successfully finding a mate is less than 1. The same situation may also be due to the effect of a species on its environment. For example, it has been suggested that blue-green algae produce a peptide which, secreted into the environment, picks up iron and thereby makes it available for some role in nitrogen fixation. Since the concentration of this peptide increases with population density, so does available nitrogen; hence the population increases (see Figure 2, p.182).

If a variable is not self-reproducing, it is usually self-damped. For instance, if a nutrient enters a lake at rate I_N (which depends on run-off) and is removed at a rate which depends on its own level and the abundance of the consumers, we have an equation of the form

$$\frac{dN}{dt} = I_N - CN$$

where C includes the activity of all the consumers. Thus

$$\frac{\partial}{\partial N}\left(\frac{dN}{dt}\right) = -C.$$

Although the nutrient level is therefore dependent on the abundances of other species and their physiological states, it is always self-damped.

The same phenomenon occurs in biochemical synthesis. Each molecular species is produced at a rate which depends on the level of its precursor and is removed at a rate which depends on its own abundance. If we recognize the different ages or developmental classes of a population as distinct variables in a system, or if we subdivide the population by location (with flow between subpopulations) or into interconvertible physiological states (e.g. starved-fed, active-inactive), then the new variables are no longer self-reproducing and have become self-damped.

In every ecological system there are some components which are not self-reproducing, such as inorganic nutrients, detritus or litter, or soil moisture. Therefore, at the lowest trophic levels of every ecosystem there are some self-damped variables. But if a model excludes these variables then their self-damping property is transferred to variables which interact with the self-damped ones. For example, consider a nutrient N which enters a system from outside, is consumed by a species X and is also removed from the system by other processes at rate C. We can represent these interactions by the equations

$$\frac{dN}{dt} = I_N - N(aX + C) \tag{3.3}$$

and

$$\frac{dX}{dt} = (bN - \theta)X \tag{3.4}$$

where a is a rate of nutrient uptake, b the rate of conversion of N into more of X and θ is the loss of species X by nutrient expenditure or death. In this system N is self-damped since $\frac{\partial}{\partial N}\left(\frac{dN}{dt}\right) = -aX - C$ but X is not necessarily self-damped. Now suppose we decide that the level of N is in equilibrium with X and that we need not recognize N as a distinct variable. Then N is replaced by $I_N/(aX + C)$, its equilibrium value in equation (3.3), and equation (3.4) becomes

$$\frac{dX}{dt} = \left(\frac{bI_N}{aX + C} - \theta\right)X,$$

so that when X is also in equilibrium

$$\frac{\partial}{\partial X}\left(\frac{dX}{dt}\right) = -\frac{abI_N X}{(ax + C)^2} .$$

Thus, in the reduced system of one variable, X has become self-damped. In general we assert that all ecosystems have self-damping variables, at least at the lower trophic level. Higher level, self-reproducing variables can be converted into self-damping variables if we remove the intermediate variables by which a species affects its own abundance.

We see, also, that there is not a unique representation of the systems we want to model. Therefore, it becomes necessary to determine when alternative models are equivalent, and when differences in modeling matter. In general, those properties which relate to equilibrium levels as functions of parameters are insensitive to the changes involved in incorporating one variable into another, while for the dynamic properties of the system greater care is required.

Once a particular representation of a system has been chosen, it can be manipulated either as a graph or a matrix. If we begin by assuming that all variables are interconnected but the interconnections differ in intensity and may not be zero, then the graph provides no information about the behavior of the system and we have to work with the distribution of interaction parameters, the coefficients of the matrix. However, if we decide that many of the possible interconnections are zero, the graph is the more informative tool. Therefore, we must be able to go back and forth between matrix and graph. A few elementary results and definitions will be required.

The determinant of a matrix can be expanded as follows. Form the product of the diagonal elements $a_{11}a_{22}a_{33}\cdots a_{nn}$. Each a_{ii} is the magnitude of a loop of length one. Now permute a pair of second subscripts and change the sign, e.g. $-a_{12}a_{21}a_{33}a_{44}\cdots a_{nn}$. The term $a_{12}a_{21}$ is the product of the a_{ij}'s around a closed loop of length two. Continue permuting second subscripts, changing the sign each time, until all $\binom{n}{2}$ terms have been generated. Now note that since each first subscript represents the input to a variable and each second subscript identifies an output, in each product we have exactly one input and one output from each variable. Therefore, each term is the product of a_{ij}'s around a simple closed loop or several disjunct loops, and each time the number of disjunct loops in a product changes, so does the sign. Finally, noting that the principal diagonal has a positive sign, we have the following result:

Let $L(m,n)$ be the product of the a_{ij}'s around m disjunct loops totaling n elements; then the determinant of order n is

$$D_n = \sum(-1)^{n+m} L(m,n) \tag{3.5}$$

where summation is over all $L(m,n)$ in the graph, for all m. A determinant of order 0, D_0, is equal to unity.

In many of the applications that follow, we make use of the notion of the feedback, F_n, of a system or a subsystem of order n; F_n is defined to be

$$F_n = \sum(-1)^{m+1} L(m,n) \tag{3.6}$$

where, again, summation is over all $L(m,n)$ in the graph, for all m. This measure has the property that if all the loops in a product are negative then the whole product contributes a negative component to the feedback. For consistency, F_0, the feedback of a system of order 0, is -1.

If, in a determinant, we substitute a column vector $K = (K_i)$ for the jth column, the resulting determinant may be expanded as before except that the terms involving a_{ij} for all i have been replaced by K_i. This means that the link from X_j to X_i in the loop has been broken, leaving only the straight path from X_i to X_j multiplied by K_i. That path is represented by $P_{ji}^{(s)}$. (We do not distinguish among the various paths from X_i to X_j except to note the number of vertices in the path by the superscript s in $P_{ji}^{(s)}$.) Each path $P_{ji}^{(s)}$ is multiplied by the product of the remaining $m-1$ loops, corresponding to the $(n-s)$-order subsystem of vertices not along the path. This subsystem is called the complement of the path and is denoted by $C(P_{ji}^{(s)})$. Therefore we have

$$D_n(K) = (-1)^n \sum K_i P_{ji}^{(s)} F_{n-s}\{C(P_{ji}^{(s)})\} \tag{3.7}$$

where $F_{n-s}\{C(P_{ji}^{(s)})\}$ is the feedback of the $(n-s)$-order system complementary to the path $P_{ji}^{(s)}$ and summation is over all paths to vertex j. Again, for consistency, a path from a variable to itself, say $P_{ii}^{(1)}$, is defined to be unity.

THE CHARACTERISTIC EQUATION AND STABILITY

IV

The local behavior of a system of differential equations

$$\frac{dX_i}{dt} = f_i(X_1,\ldots,X_n) \qquad (i=1,\ldots,n)$$

near an equilibrium point depends on the roots (eigenvalues) of the characteristic equation

$$|A - \lambda I| = 0 \tag{4.1}$$

where $A = (a_{ij})$ is the matrix of first partial derivatives $\frac{\partial f_i}{\partial X_j}$ evaluated at the equilibrium point.

If the real parts of all the roots are negative, the system returns to equilibrium after a small perturbation. If the real parts of all the roots are positive, the system moves away from equilibrium (is locally unstable). If some roots have positive and some negative real parts, the behavior of the system depends on how it is perturbed; it sometimes returns to equilibrium but for other displacements moves away. In biological systems we usually assume the perturbations to be unconstrained so that eventually the system will be displaced in a direction which allows the positive root to lead the system away from equilibrium. A single zero real part gives a neutral or passive equilibrium, but multiple zero roots can give unbounded solutions (unstable equilibrium). If a root is complex the system oscillates at a frequency given by the imaginary part while the amplitude behaves according to the real part of the root.

It must be noted, however, that regardless of whether a system is stable or unstable, the rate of change near an equilibrium point is small, and real systems may therefore be found near stable or unstable points. In the long run the system moves away from an unstable equilibrium, but parameter changes or extraneous perturbations may leave it near such a point (usually a saddle point) and it will then remain there for a long time.

To analyze equation (4.1) we expand it as the polynomial

$$|A - \lambda I| = \lambda^n + \sum_{k=1}^{n} (-1)^k \lambda^{n-k} (\sum D_k) \tag{4.2}$$

where $\sum D_k$ is the sum of the determinants of all subsystems of order k. Now by (3.5) and (3.6) which relate D_n to its loop structure and feedback, we obtain the characteristic polynomial

$$P(\lambda) = \lambda^n - \sum_{k=0}^{n-1} \lambda^k F_{n-k}. \tag{4.3}$$

Now we can use well known results concerning the roots and coefficients of equations to relate the local dynamics of a system to the structure of its graph.

First, the coefficient of λ^{n-k} is $(-1)^k$ times the sum of the products of the roots taken k at a time. Thus, when k equals unity, we have

$$\sum \lambda_i = F_1 = \sum a_{ii}.$$

Since the arithmetic average of the λ_i's is the arithmetic average of the diagonal elements of A, any change in the off-diagonal elements a_{ij} simply redistributes the λ_i's around the same mean. To find the spread of the λ_i's let us calculate the sum of squared deviations from the average, $\bar{\lambda}$, within a given matrix. Using equation (3.7) and the relation

$$-\sum_{i \neq j} \lambda_i \lambda_j = -\sum_{i \neq j} D_{ij} = F_2$$

where

$$D_{ij} = \begin{vmatrix} a_{ii} & a_{ij} \\ a_{ji} & a_{jj} \end{vmatrix} ,$$

it follows that

$$\sum \lambda_i^2 = \sum a_{ii}^2 + 2 \sum_{i \neq j} a_{ij} a_{ji} .$$

Hence

$$\sum \lambda_i^2 - n\bar{\lambda}^2 = \sum a_{ii}^2 - 1/n(\sum a_{ii})^2 + 2 \sum_{i \neq j} a_{ij} a_{ji}$$

$$= \{\sum a_{ii}^2 - 1/n(\sum a_{ii})^2\} + n(n-1)\overline{a_{ij} a_{ji}} \qquad (4.4)$$

where the term in brace brackets represents the spread of the diagonal elements of A and the bar indicates arithmetic average. Two observations follow: first, if the average loop of length 2 $(\overline{a_{ij} a_{ji}})$ is negative then for large enough n the spread of λ will be negative. Hence at least some of the λ_i's will be complex and the system will necessarily oscillate.

Second, this effect is weakened if the spread of the diagonal elements of A is large, but if all the a_{ii}'s are the same then any negative value of $\overline{a_{ij} a_{ji}}$ is sufficient to cause oscillation. This is a weak illustration of the general principal that if systems change at very different rates when uncoupled (large spread in the a_{ii}'s) then in the whole system they will still behave more or less independently.

Now divide the characteristic equation by $-\lambda^n F_n$. This gives

$$\left(\frac{1}{\lambda}\right)^n + \frac{F_{n-1}}{F_n}\left(\frac{1}{\lambda}\right)^{n-1} + \frac{F_{n-2}}{F_n}\left(\frac{1}{\lambda}\right)^{n-2} + \ldots + \frac{F_1}{F_n}\left(\frac{1}{\lambda}\right) - \frac{1}{F_n} = 0.$$

This equation in $\frac{1}{\lambda}$ may be treated in the same way. Since

$$\sum \frac{1}{\lambda_i} = \frac{-F_{n-1}}{F_n} ,$$

$H(\lambda)$, the harmonic mean of the characteristic roots of A, is given by

$$H(\lambda) = \frac{-nF_n}{F_{n-1}} .$$

But if all the λ_i's are of the same sign, real, and not identical, the arithmetic mean exceeds the harmonic mean. If this relation does not hold, i.e. if

$$\frac{F_1}{n} < -n \frac{F_n}{F_{n-1}} , \qquad (4.5)$$

we can conclude that either the system has complex roots or roots of different signs. Similar relations hold for other coefficients. Clearly,

$$-F_k = (-1)^k \sum \lambda_{i_1} \lambda_{i_2} \cdots \lambda_{i_k}$$

and

$$\frac{-F_{n-k}}{F_n} = (-1)^k \sum \left(\frac{1}{\lambda_{i_1}}\right) \left(\frac{1}{\lambda_{i_2}}\right) \left(\frac{1}{\lambda_{i_3}}\right) \cdots \left(\frac{1}{\lambda_{i_k}}\right) .$$

Since each of these products contains $\binom{n}{k}$ terms, inequality (4.5) becomes

$$\frac{-F_k}{\binom{n}{k}} < \binom{n}{k} \frac{F_n}{F_{n-k}}$$

or

$$F_k F_{n-k} < -\binom{n}{k}^2 F_n .$$

Local stability requires that all λ_i's have negative real parts. This, in turn, imposes two requirements on the coefficients F_k. First, for all k, F_k must be negative. This is sufficient to exclude positive real roots so that in a symmetric matrix it is a sufficient test of stability. Note that the requirement that feedback be negative at each level means that the addition of loops of length greater than k cannot stabilize a system which is unstable due to positive feedback at level k.

The second requirement is expressed as a series of inequalities (the Routh-Hurwitz criteria) of which the first is

$$F_1F_2 + F_3 > 0.$$

This may be expanded in terms of loops to give

$$\sum(-L_i^2L_j - 2L_iL_jL_k + L_iL_{ij} + L_{ijk}) > 0$$

where the subscripts identify vertices in a loop. The first three terms will be positive when the individual loops are negative, while the last term refers to loops of length three and will be negative.

Combining these two requirements, we can state that a system is stable if feedback is negative at all levels, but higher level negative feedback must not be too strong compared to lower level negative feedback. This latter requirement is necessary because the length of a loop is analogous to time lag, and a long lag with strong negative response can throw a system into increasing oscillations.

In Figure 2 we show some examples of graphs which can be seen to be unstable by inspection.

If the graph of a system is given, we can either determine the local stability unambiguously or decide what has to be measured to determine whether it is stable or not. Thus, in Figure 2(d) the system will certainly be oscillatory unstable if

$$a_{11}a_{12} < -a_{32}a_{13}. \tag{4.6}$$

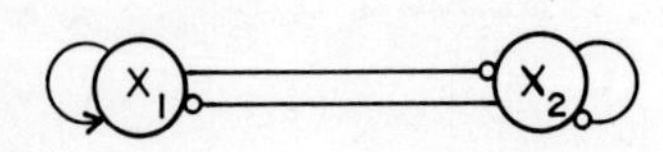

(a) F_1 is of undetermined sign but F_2 is positive.

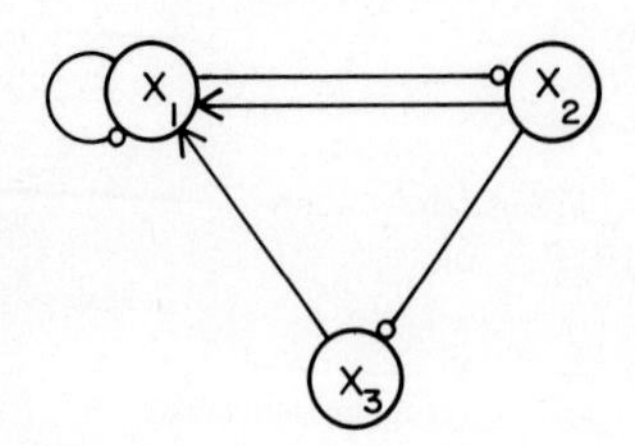

(b) F_1 and F_2 are negative but F_3 is positive.

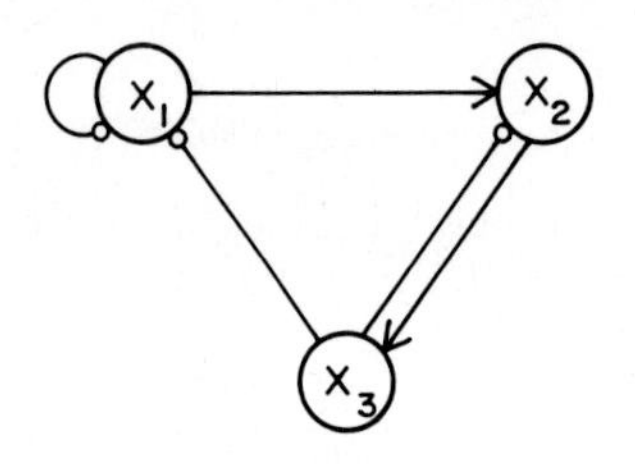

(c) F_1,F_2 and F_3 are negative but the first Routh-Hurwitz inequality is violated.

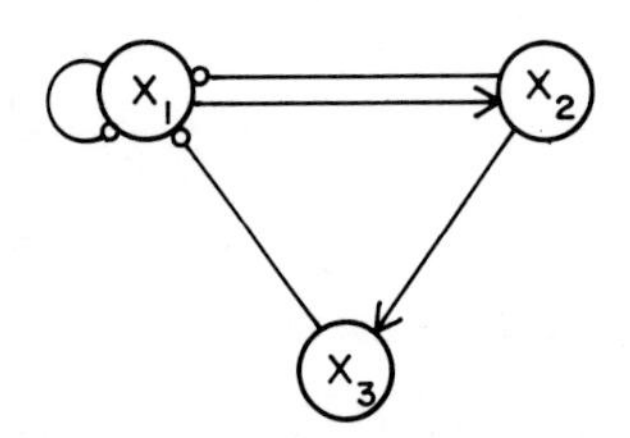

(d) The loop of length 2 has been relocated.

Figure 2

Another class of problems arises if we do not know the exact structure of the graph but only some rule for its construction. For example, suppose that in a system of n variables a fraction p_1 are self-damped with average intensity a_{ii} and variance σ^2, that the probability of a link between any two variables is p, that if they are linked the interaction is reciprocal with the average $a_{ij} = \bar{a}$ but with a_{ij} and a_{ji} correlated, and all other interactions are independent. Then condition (4.6) becomes

$$-n(n-1)p_1^{\,2}(a_{ii}+\sigma^2) - n(n-1)p_1 p a_{ii}\{\bar{a}^2 + \mathrm{cov}(a_{ij},a_{ji})\} - \frac{n(n-1)(n-2)}{3}(p_1^{\,3}a_{ii}^{\,3} - p^3\bar{a}^3) > 0$$

Thus, large values of self-damping (p_1 and a_{ii}) stabilize the system, as does a large variance in a_{ii} (this is really a partial uncoupling of the system), and high connectivity (p) and interaction intensity $\bar{a}$ destabilize the system. As n increases, the terms involving $n(n-1)(n-2)$ dominate and the condition approaches

$$-p_1^{\,3}a_{ii}^{\,3} + p^3\bar{a}^3 > 0. \qquad (4.7)$$

This, of course, is only a one-sided test. If the inequality is violated the system is unstable, but if (4.7) is satisfied, other stability criteria may still be violated.

A second model examines the requirement that $F_k < 0$ for all k. Suppose, for simplicity, that $a_{ii} = -1$ for all i, that the average a_{ij}

is zero, and that all a_{ij} are independent except that the average value

$$E(a_{ij}a_{ji}) = \mathrm{cov}(a_{ij},a_{ji})$$

is not zero. Then we can express the average feedback of an n-variable system in terms of the feedbacks of smaller systems, viz.

$$\overline{F}_n = \overline{F}_{n-1} - \mathrm{cov}(a_{ij},a_{ji})\overline{F}_{n-2},$$

$$F_0 = -1,$$

$$F_1 = -1.$$

If the covariance of a_{ij} and a_{ji} is positive (as in a competitive interaction or symbiosis) then $\overline{F}_n$ can only increase with n. Therefore when $\mathrm{cov}(a_{ij},a_{ji}) > 1$, $\overline{F}_n$ will be positive. This does not prove that the system of order n is necessarily unstable but only that the average feedback will be positive, so that, roughly speaking, the probability of stability is decreasing with n. The assumption that average a_{ij} is zero is made only for convenience. Similar recurrence relations can be obtained under other assumptions.

Models of this kind require mathematically tractable assumptions about the structure of the graphs; however, we do not know how the graphs of real systems depart from the models. But insofar as we show that graph properties determine system behavior we can then pose experimental problems. For example, we can askhow the number of long loops in real systems deviates from the number expected from combinatorial analysis and how this deviation affects the dynamics? Because of the obvious power of graphical analysis it is important to emphasize that the model is no substitute for experiment; rather, the model identifies new items of experimental interest such as connectivity or departure from randomness.

The effect of departures from randomness may be examined by differentiating the characteristic root with respect to feedback at any level. Differentiating (4.3) with respect to F_k we get

$$\frac{d\lambda}{dF_k} = \frac{\lambda^{n-k}}{\frac{\partial P(\lambda)}{\partial \lambda}} .$$

The denominator on the right-hand side is the derivative of the characteristic polynomial at the given root λ and is therefore independent of k. Thus the sensitivity of λ to a change in F_k is proportional to λ^{n-k}. It follows that as $|\lambda|$ increases, so does the relative importance of low-order feedback (large exponents of λ) while if $|\lambda|$ is small the higher-order feedbacks increase in relative significance; as $\lambda \to 0$ the influence of F_n is most important. This result is intuitively reasonable; short loops predominate in determining short term behavior and long loops are important for long term behavior.

MOVING EQUILIBRIA

V

Now we examine the question of how the equilibrium level of a variable changes when some parameter of the system changes. In the first instance the parameter change does not depend on the state variables but rather on other processes external to the model. However, it can also happen that the parameters are themselves responding slowly to the variables and we then have the problem of coupled fast and slow systems.

The parameters may be properties of the external environment which enter into the equations for the change of all variables, or they may be genetic properties of the interacting species which alter in response to natural selection. If there is a genetically determined parameter affecting the fecunditv of heat tolerance of a species it apnears only in the equation for that species. But if the parameter affects the vulnerability to predation it appears in the equations for both predator and prey, and may be altered by the evolution of both.

Return now to equation (3.1) and allow the system to be at equilibrium. Then, differentiating with respect to some parameter C_h we obtain the equation

$$\frac{\partial f_i}{\partial C_h} + \sum \frac{\partial f_i}{\partial X_j} \frac{\partial \hat{X}_j}{\partial C_h} = 0.$$

The derivatives $\frac{\partial f_i}{\partial X_j}$ are the familiar a_{ij}'s introduced in (3.2). Therefore we have a new set of linear equations in the unknown quantities $\frac{\partial \hat{X}_i}{\partial C_h}$, viz.

$$\sum_j a_{ij} \frac{\partial \hat{X}_j}{\partial C_h} = - \frac{\partial f_i}{\partial C_h}, \quad (i = 1,\ldots,n).$$

The solution for $\frac{\partial \hat{X}_i}{\partial C_h}$ is

$$\frac{\partial \hat{X}_i}{\partial C_h} = \frac{\begin{vmatrix} a_{11} \cdots -\frac{\partial f_1}{\partial C_h} \cdots a_{1n} \\ \vdots \qquad \vdots \qquad \vdots \\ a_{n1} \cdots -\frac{\partial f_n}{\partial C_h} \cdots a_{nn} \end{vmatrix}}{|A|}, \quad (i = 1,\ldots,n) \tag{5.1}$$

that is, the column vector $\left(\frac{\partial f_j}{\partial C_h}\right)$ is substituted for the ith column in the determinant of A, and this is divided by the determinant of the whole system. From equations (3.5) and (3.6) we see that the determinant of order n is

$$D_n = (-1)^{n+1} F_n .$$

In the numerator of (5.1) the replacement of each a_{ji} in the ith column by $-\frac{\partial f_j}{\partial C_h}$ breaks a loop from X_j to X_i and back to X_j, leaving the straight path from X_j to X_i multiplied by $-\frac{\partial f_j}{\partial C_h}$. Now, a term which previously had m loops has m-1 loops, but in terms of feedback it is still multiplied by $(-1)^m$. The negative sign preceding $\frac{\partial f_j}{\partial C_h}$ makes this $(-1)^{m-1}$, so that finally

$$\frac{\partial \hat{X}_i}{\partial C_h} = \frac{\sum \frac{\partial f_j}{\partial C_h} P_{ij}^{(k)} F_{n-k}\{C[P_{ij}^{(k)}]\}}{F_n}, \quad (i = 1,\ldots,n)$$

where $P_{ij}^{(k)}$ is a simple path (one that does not cross itself) from X_j to X_i which includes k vertices, $C[P_{ij}^{(k)}]$ is the complement of the path (that is, the subsystem of all variables not on the path), $F_{n-k}\{C[P_{ij}^{(k)}$ is the feedback of this complement, and the summation is over all paths from all variables to X_i.

If we assume that the system is locally stable, then F_n is negative. Therefore if $F_{n-k}\{C[P_{ij}^{(k)}]\}$ is also negative the change in X_i due to a change in C_h has the same sign as the product of the a_{ij}'s along the path multiplied by the sign of the input $\frac{\partial f_j}{\partial C_h}$. In this case the results are more or less what we would have expected without the graph. But if the complement has zero feedback then the path has no effect, and if the complement has positive feedback the result is the opposite of what we would have expected.

In Figure 3 we return to the graph of Figure 1(a) to show the decomposition into path and complement when an increased rate of input enters at either N or P. The system may be interpreted as a lake in which N and P are nitrate and phosphate, while B and G represent blue-green and green algae. Whereas green algae require both nutrients, blue-greens need phosphate but fix nitrogen, and inhibit the growth of green algae. Figure 3(a) shows the whole system, while Figures 3(b)-(e) show the response of each variable to nitrate increment; the path is on the left side of the figure while the complement is on the right, followed by the sign of the effect. Recall that there is negative denominator, F_n, for each term and $P_{ii}^{(1)} = 1$, $F_0 = -1$. Therefore the sign is (-1) times path times complement. In Figure 3(b) the complement of N is the positive loop (P,B,G,P). Therefore, increasing the input to N reduces the equilibrium level of N. This comes about biologically because adding N increases the growth of G, which competes for phosphate with B, thus reducing B and removing some of the inhibition of B on G, thereby increasing even further the consumption of

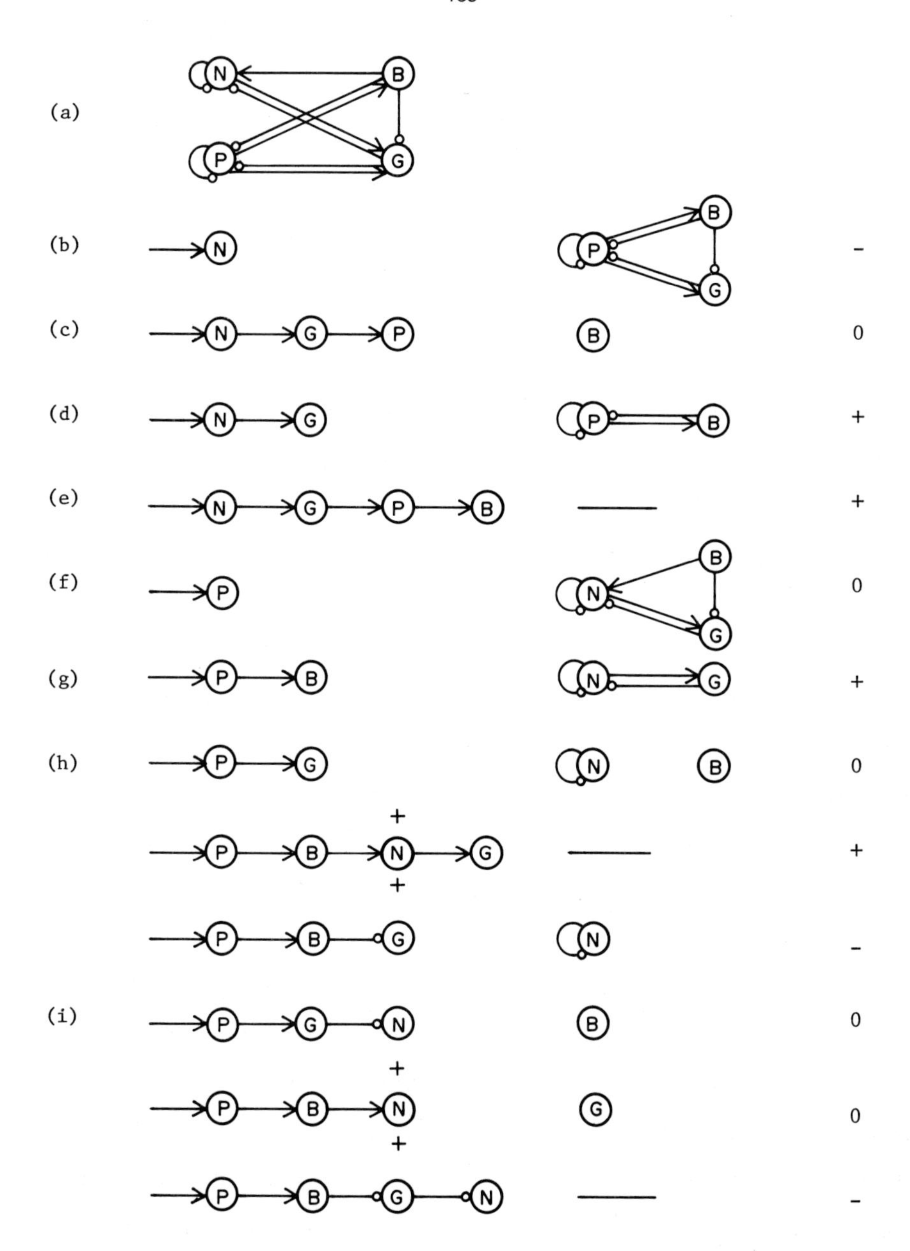
(a)
N
B
P
G
(b)
N
P
B
G
-
(c)
N
G
P
B
0
(d)
N
G
P
B
+
(e)
N
G
P
B
+
(f)
P
N
B
G
0
(g)
P
B
N
G
+
(h)
P
G
N
B
0
+
P
B
N
G
+
+
P
B
G
N
-
(i)
P
G
N
B
0
+
P
B
N
G
0
+
P
B
G
N
-

Figure 3

N by G. This effect disappears without the negative link from B to G.

In Figures 3(c),(f),(h) and (i) there are paths which have a zero complement and therefore no effect. In Figures 3(h),(i) there is more than one path from P to the target variable. In the latter case all paths but one have zero complements but in Figure 3(h) two paths have opposite effects. Therefore, the graph alone cannot determine the sign of the effect of P on G; measurements are necessary, but we can see that the effect will be negative or positive depending on the relative strengths of the inhibition and nitrogen fixation links.

In part VIII this method of analysis will be applied to a series of problems.

CHARACTERISTIC VECTORS

VI

The general solution of a set of linear differential equations expresses each variable in the form

$$X_i = \sum_j D_{ij} e^{\lambda_j t}$$

where the λ_j's are the distinct characteristic roots of the matrix of coefficients. (We ignore here the case of multiple roots.) The constants D_{ij} depend on initial conditions so that some of them may be zero, but each X_i is usually a function of most of the roots. However, there exists a linear combination of the variables, the characteristic vector associated with each characteristic root, which satisfies the relation

$$\frac{dV_i}{dt} = \lambda_i V_i, \quad (i = 1,\ldots,n)$$

where

$$V_i = \sum C_{ij} X_j$$

and the coefficients C_{ij} are unique to a constant multiplier. If λ_i is a large negative root then V_i returns rapidly to equilibrium, and is therefore a relatively stable descriptor of the system. But for small λ_i the rate of return is slow. Since the average of the λ_i's is the average of the diagonal elements (self-damping) in the matrix of coefficients, any change in the off-diagonal elements of the matrix results in a redistribution of λ_i's, increasing some and reducing others. Thus, as a system changes because of new variables, new interconnections, or altered parameters, some measures appear which are increasingly stable while others become less stable. Our problem is to interpret these measures, the coefficients of the characteristic vector.

To find the coefficient C_i of X_i in a given characteristic vector, choose any row of the matrix

$$(A - \lambda I) = \begin{bmatrix} a_{11}-\lambda & a_{12} & \cdots & a_{1n} \\ a_{21} & a_{22}-\lambda & \cdots & a_{2n} \\ \vdots & & & \vdots \\ a_{n1} & a_{n2} & \cdots & a_{nn}-\lambda \end{bmatrix}.$$

The cofactor of a_{ji} in an expansion by row j is the coefficient of X_i. But the cofactor of a_{ji} is the product along all paths from X_j to X_i each multiplied by the feedback of its complement in the matrix. If we expand C_i in powers of λ we get

$$C_i = \sum_k (-1)^k \lambda^k \sum P_{ij}{}^{(h)} F_{n-k-h}\{\text{Comp.}\}.$$

Here $F_{n-k-h}\{\text{Comp.}\}$ is the feedback of the complement of the path from X_j to X_i in subsystems of order $(n-k-h)$, and since summation is over all h and all such paths, $\sum P_{ij}{}^{(h)} F_{n-k-h}\{\text{Comp.}\}$ is equivalent to the sensitivity of the equilibrium level $\hat{X}_i$ to a unit change entering the system at X_j in subsystems of size k. Let $S_{ij}{}^{(n-k)}$ be the sensitivity of X_i to X_j in all subsystems of $(n-k)$ variables; then

$$C_i = \sum_{k=0}^{n-1} (-1)^k \lambda^k S_{ij}{}^{(n-k)}.$$

Thus C_i, the coefficient of X_i in a given characteristic vector is a linear combination, weighted by powers of λ, of the sensitivity of the variable to changes in a given standard variable in subsystems of different sizes. If λ is very large, C_i is dominated by large powers of λ and therefore two variables will have similar coefficients if their sensitivities are similar in small subsystems. But for small λ the sensitivities in larger subsystems become more important.

The choice of a standard X_j to measure sensitivities introduces an asymmetry in the form of the expressions. But the way in which the coefficients depend on the structure can be seen by comparing the coefficients of two variables. Consider the system in Figure 4(a).

First, in order to compare the coefficients of variables N and H, which have opposite sensitivities to A_1, we take A_1 as the standard and express its cofactors as sensitivities. The coefficients of N,H are

$$C_N = \underset{(-)}{\lambda^2 a_{NA_1}} + \underset{(+)}{a_{HA_1} a_{A_2H} a_{NA_2}} - \underset{(-)}{a_{NA_1} a_{HA_2} a_{A_2H}} , \tag{6.1}$$

$$C_H = \underset{(+)}{\lambda^2 a_{HA_2}} - \underset{(+)}{a_{HA_1} a_{A_2N} a_{NA_2}} + \underset{(-)}{a_{NA_1} a_{A_2N} a_{HA_2}} . \tag{6.2}$$

The sign in parentheses under each term indicates its sign. Note that for large λ, N and H are likely to have opposite signs. For small λ the sign of the constant terms must be determined. Factoring out a_{A_2H} in equation (6.1) we find the constant term has the sign of

$$\frac{a_{HA_1}}{a_{NA_1}} + \frac{a_{HA_2}}{a_{NA_2}} .$$

Since the same relation holds in (6.8), C_N and C_H have the same sign in eigenvectors corresponding to very small eigenvalues λ_i.

To compare the coefficients of variables on the same level, take N as the standard; then

$$C_{A_1} = \underset{(+)}{\lambda^2 a_{A_1N}} + \underset{(-)}{a_{A_2N} a_{HA_2} a_{A_1H}} - \underset{(+)}{a_{A_1N} a_{A_2H} a_{HA_2}} ,$$

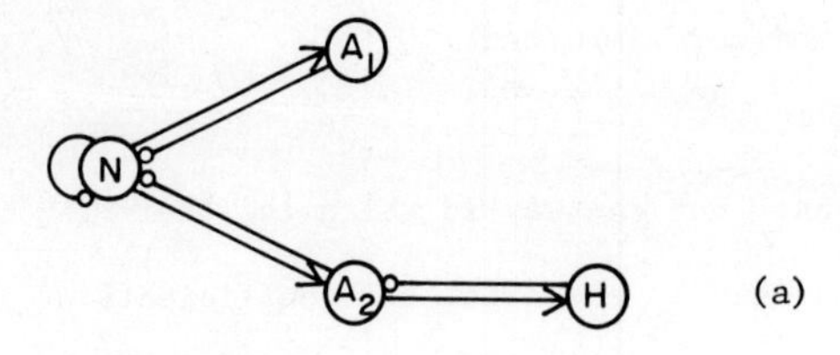

(a)

	N	A_1	A_2	H
N	0	+	0	0
A_1	-	+	0	-
A_2	0	0	0	+
H	0	+	-	0

One of two competitors for a common resource is itself preyed upon; the other is resistant.

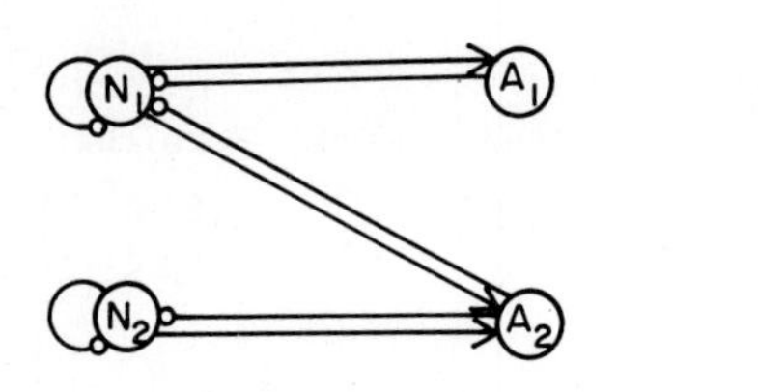

(b)

	N_1	N_2	A_1	A_2
N_1	0	0	+	0
N_2	0	0	-	+
A_1	-	+	+	-
A_2	0	-	-	+

One of two competitors, A_1, specializes in their common resource N_1 while A_2 has an additional alternative resource or requirement.

Figure 4

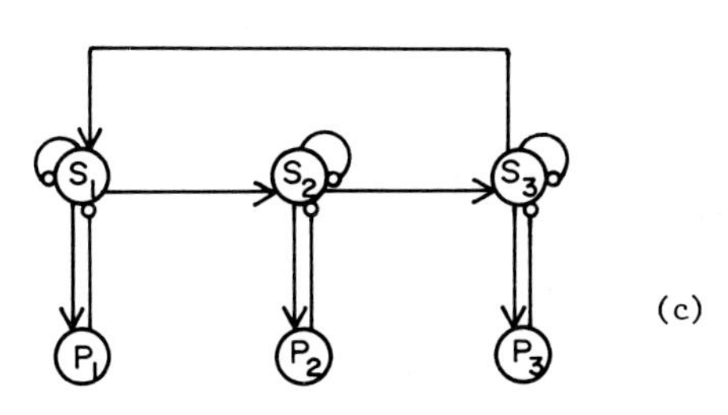

(c)

	S_1	S_2	S_3	P_1	P_2	P_3
S_1	0	0	0	+	0	0
S_2	0	0	0	0	+	0
S_3	0	0	0	0	0	+
P_1	-	0	0	+	-	0
P_2	0	-	0	0	+	-
P_3	0	0	-	-	0	+

The three predators or parasitoids P_1,P_2,P_3 attack different stages of the life cycle of the same host.

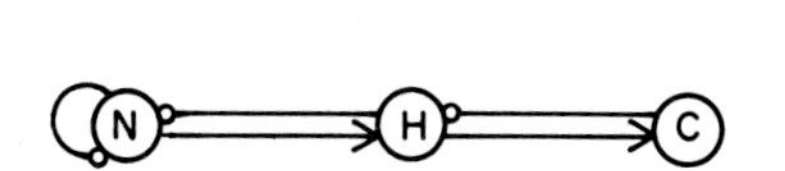

(d)

	N	H	C
N	+	0	+
H	0	0	+
C	+	-	+

A simple three-level system of nutrient (plant), herbivore and carnivore.

Figure 4

$$C_{A_2} = \lambda^2 a_{A_2N} + a_{A_1N}a_{HA_1}a_{A_2H} - a_{A_2N}a_{A_1H}a_{HA_1} .$$
$$\quad (+) \qquad\qquad (-) \qquad\qquad (+)$$

For large λ the term in λ^2 dominates and C_{A_2} and C_{A_1} have the same sign. For small λ we have to examine the constant term (provided it is not zero). C_{A_1} will have the sign of

$$\frac{a_{A_2N}}{a_{A_1N}} - \frac{a_{A_2H}}{a_{A_1H}}$$

and C_{A_2} will have the opposite sign. Finally, to compare the coefficients of A_1 and H_1, take N as the standard; then

$$C_{A_1} = \lambda^2 a_{A_1N} + a_{A_2N}a_{HA_2}a_{A_1H} - a_{A_1N}a_{A_2H}a_{HA_2} ,$$
$$\quad (+) \qquad\qquad (-) \qquad\qquad (+)$$

$$C_H = -\lambda[a_{A_1N}a_{HA_1} + a_{A_2N}a_{HA_2}] .$$
$$(+)$$

Thus, for large λ the coefficients of A_1, A_2, H will have the same sign and the coefficient of N will have the opposite sign. Hence the largest characteristic vector compares nutrient to total consumers (direct and indirect). For small λ, C_H will have the same sign as one of A_1, A_2. Therefore the smallest characteristic vector reflects the difference between the top and bottom layers, with A_1 and A_2 divided between them.

If A_1 and A_2 are biologically equivalent (that is, are identical in their interactions with N and H) then the system of equations has one zero eigenvalue. In that case, we can lump A_1 and A_2 in a new variable A and the system reduces to

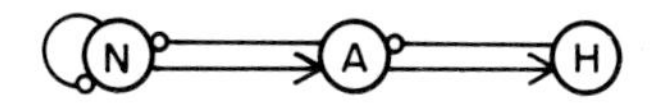

The subject of lumping will be discussed in the following section.

In a community of competitors all the interaction coefficients are of the same sign. In this case it can be shown (Frobenius's theorem) that the vector corresponding to the largest eigenvalue has all positive coefficients; this may be interpreted as indicating that the coefficients of λ^{n-1} are all the a_{ij}'s, which have the same sign.

The chief result of this section, then, is that characteristic vectors can be interpreted as measures of community structure, and that as the characteristic roots spread out about their arithmetic mean some of these measures become strongly stabilized while others become less stable. However, the biological interpretation of characteristic vectors has barely begun.

SOME SIMPLIFICATIONS

VII

In this section we indicate briefly a few rules for simplifying systems.

Intermediate Variables

It is always possible to insert a variable along a path between any two variables. For instance, we can divide a species into any number of age classes, or a gradient into arbitrarily many compartments. The question is, does it really matter? Does it affect the characteristic roots of the system or the response to parameter change?

Consider an arbitrary system in which we have a link from X_1 to X_2. The characteristic polynomial can be written in the form

$$P(\lambda) = C_1(\lambda) + a_{21}C_2(\lambda)$$

where $C_1(\lambda)$ consists of all decompositions into loops which do not include the $X_1 - X_2$ link, while $C_2(\lambda)$ is the cofactor of a_{21} in the characteristic matrix.

Now introduce a new variable Y between X_1 and X_2; clearly

$$\frac{dY}{dt} = a_{y1}X_1 - a_{yy}Y.$$

Thus, at an equilibrium point

$$\hat{Y} = \frac{a_{y1}X_1}{a_{yy}}. \tag{7.1}$$

Consistency requires that X_2 is formed at the same rate as before, so that

$$a_{2y}Y = a_{21}X_1. \tag{7.2}$$

The characteristic equation is then

$$(a_{yy}-\lambda)C_1(\lambda) + a_{2y}a_{y1}C_2(\lambda) = 0.$$

Now combining (7.1) and (7.2) we obtain

$$a_{2y}a_{y1} = a_{21}a_{yy}.$$

Thus the characteristic equation is

$$(a_{yy}-\lambda)C_1(\lambda) + a_{21}a_{yy}C_2(\lambda) = 0. \tag{7.3}$$

Divide (7.3) by a_{yy}; then

$$(1 - \frac{\lambda}{a_{yy}})C_1(\lambda) + a_{21}C_2(\lambda) = 0.$$

If y is a short-lived intermediary, a_{yy} will be large compared to any root λ of the original system, so that near those roots

$$P(\lambda) \approx C_1(\lambda) + a_{21}C_2(\lambda)$$

and the system with Y will have approximately the same first n roots.

Similar Variables

It often happens that the variables of a system fall into functional subsets, e.g. predators. A particularly simple situation arises if X_1 and X_2 interact symmetrically with each other and indentically with all other variables; then

$$a_{11} = a_{22}, \quad a_{21} = a_{12}, \quad a_{i1} = a_{i2} \quad i = 3,\dots,n, \quad a_{1j} = a_{2j} \quad j = 3,\dots,n.$$

The characteristic equation of this system is

$$\begin{vmatrix} a_{11}-\lambda & a_{12} & a_{13} & a_{14} & \cdots & a_{1n} \\ a_{21} & a_{22}-\lambda & a_{13} & a_{14} & \cdots & a_{1n} \\ a_{31} & a_{31} & & & & \\ a_{41} & a_{41} & & & & \\ \vdots & \vdots & & & & \vdots \\ a_{n1} & a_{n1} & & & \cdots & a_{nn}-\lambda \end{vmatrix} = 0.$$

If we subtract the second row of the above determinant from the first, then add the first column to the second, we obtain the equation

$$\begin{vmatrix} a_{11}-a_{21}-\lambda & 0 & & & \cdots & 0 \\ a_{21} & a_{22}+a_{21}-\lambda & a_{13} & a_{14} & \cdots & a_{1n} \\ a_{31} & 2a_{31} & & & & \\ a_{41} & 2a_{41} & & & & \\ \vdots & \vdots & & & & \vdots \\ a_{n1} & 2a_{n1} & & & \cdots & a_{nn}-\lambda \end{vmatrix} = 0.$$

When expanded by the first row of the matrix this determinantal equation becomes

$$(a_{11}-a_{21}-\lambda)\begin{vmatrix} a_{22}+a_{21}-\lambda & a_{13} & a_{14} & \cdots & a_{1n} \\ 2a_{31} & & & & \\ \vdots & & & & \vdots \\ 2a_{n1} & & & \cdots & a_{nn}-\lambda \end{vmatrix} = 0.$$

Now look at the subsystem of X_1, X_2. Its characteristic equation is

$$(a_{11}-\lambda)^2 - a_{12}^{\ 2} = 0$$

so that

$$\lambda = a_{11} \pm a_{12}.$$

Corresponding to the eigenvalue $a_{11} + a_{12}$ is the characteristic vector $X_1 + X_2$, while the vector $X_1 - X_2$ corresponds to the root $a_{11} - a_{12}$. Therefore the characteristic equation consists of two parts; the factor $a_{11} - a_{12} - \lambda$ determines the root associated with the relative abundances of X_1 and X_2, a matter of indifference to the rest of the community. In the rest of the equation, the two variables are lumped and are represented by their sum $X_1 + X_2$, which has the self-damping term $a_{11} + a_{12}$.

Homogeneous Linear Subsystems

In many of the applications to the study of moving equilibria, it is important to find the sign of the feedback of a complementary subsystem. If that subsystem is made up of variables representing age or stage classes, positions in a gradient, or reversible states, then the equations will often be linear in those variables (it is not necessary that the subsystem be linear in all variables of the whole) or else of the form

$$\frac{dX_i}{dt} = X_i f_i(X_1, X_2, \ldots, X_k)$$

where f_i is linear and homogeneous in the k variables of the subsystem. Then at an equilibrium point we have a system of k linear, homogeneous equations to solve. In order for there to be a non-trivial solution, the determinant of the system must vanish; hence the subsystem has zero feedback.

RESPONSES TO PARAMETER CHANGE

VIII

The results of graphical analyses may be used in several ways. First, assume that the parameters are environmental factors which are directly observable. Then there is a simple, straightforward comparison between predicted and observed variation in space or time (however, the temporal variation in parameters must be slow enough for the moving equilibrium model to be applicable).

On the other hand we may know the graph and the observed geographic pattern, from which predictions can be made about which aspects of the environment are varying. For instance, in Figure 4(a) we see that A_2 changes only if the environmental variation enters by way of the herbivore H. If most of the environmental differences affecting the community enter by way of conditions that affect H directly, we would expect H and N to be relatively constant from place to place, and A_1 and A_2 to show comparable variation with negative correlation. But if the input of nutrient varies, this generates changes in A_1 which will be uncorrelated with A_2. Thus, we can go from the graph and table to test predictions about environment. The probability of confirming all predictions by chance would be 2^{-n^2}. For n greater than three this is already significant at less than the .002 level so that the qualitative model has strong testability. In practice it will not be possible to find or produce inputs to each variable, and some of the predictions may be ambiguous. Then the strength of the test is correspondingly reduced.

A more general use can also be made of these graphs. Without being able to measure the environments where the organisms are observed we may be able to determine that a particular factor is more variable in one region than in another. If inputs to N (cf. Figure 4(a)) or N_1 (cf. Figure 4(b)) are more variable, this should increase the variation in A_1 in the region but leave the other variables unaffected.

An alternative interpretation of the parameters is in terms of a genetic model in which each parameter depends on the genotypes of the species of the community. The argument proceeds as follows. At demographic equilibrium,

$$\frac{dX_i}{dt} = f_i(X_1, X_2, \ldots; C_1, C_2, \ldots) \qquad (i = 1, \ldots, n)$$

is equal to zero. Then, if C_h is a genetically determined parameter, if $\frac{\partial f_i}{\partial C_h} > 0$, and if genetic variation arises for C_h, then C_h will increase. This is a straightforward consequence of Mendelian population genetics which is independent of the community structure. But the consequences of this genetic change do depend on the community. First, suppose that C_h affects the viability or fecundity of species A_1 in Figures 4(a) or 4(b). Then in other cases, A_1 itself will increase and some of the other variables in the system will also be altered. But if the genetic change arises in A_2 of Figure 4(a), the level of A_2 will not alter. Any increased fecundity or viability will affect H instead. And in Figure 4(c), any evolution of viability or fecundity in species H will affect C but H itself will not increase. (Increased viability means that a larger proportion of H is eaten by C than previously had been the case, fewer of H are killed by the physical factors to which the population adapted, and since C increased, the selection with H produced a deterioration of its environment.)

A different situation arises if the gene selected acts on the rate of predation. Then, regardless of which species the genetic change arises in, it affects predator and prey with opposite sign. For example, suppose that

A_2 in Figure 4(a) becomes more resistant to H. The positive input to A_2 increases H but has no effect on A_2. But the same genetic change appears also as a negative input to H resulting in an increase in A_2 and a decrease in A_1. The final result is that both A_2 and H increase. The increase in A_2 provides more food for H even though A_2 is harder to catch, while the increase in A_2 takes a larger share of nutrient and reduces A_1.

However, these two kinds of genetic changes are not independent. Suppose that the gene in question affects an allocation of energy or other resource between the production of tough or spiney leaves and seed production. The allele which favors seed production would then have a selective value proportional to $(S - bH)$ where S is the advantage with regard to seed set and b the advantage with respect to predator avoidance. If this term is positive, that is, if the increased reproduction outweighs the increased predation loss, the gene will be selected for, resulting in a decrease in the population. Note that what is selected depends on the level of H. An increase in reproduction enters as a positive input to the plant, which increases H, and also as a direct positive input to H due to easier harvesting. Therefore evolution of the plant will result in a decrease in its abundance, but also in an increase in H which may eventually stop that selection (when $S = bH$). Other cases are considered in Levins (1976). The general principle is that equations for the course of Mendelian selection and those which determine demographic changes are distinct, that the results of selection do not determine the consequences of that selection, and therefore that there is no necessary relation between the evolution of species and the evolution of communities.

REFERENCES

Levins, R. (1976) Evolution in communities near equilibrium. In *Ecology and Evolution*, M. Cody and J. Diamond, eds. (Cambridge: Harvard University Press).

Mason, S.J. (1953) Feedback theory - some properties of signal-flow graphs. *Proceedings of the Institute of Radio Engineers, 41*, 1144-1156.

Roberts, F.S. (1976) *Discrete Mathematical Models*. Englewood Cliffs, N.J.: Prentice-Hall, Inc.

Mathematical Models in Population Biology

Professor J. Maynard Smith

School of Biological Sciences
University of Sussex
Falmer, Brighton
Sussex, England

EVOLUTION AND THE THEORY OF GAMES

1. The Biological Problem

In the Hamadryas baboon, males are substantially larger than females. A troop of baboons is subdivided into a number of 'one-male groups', consisting of one adult male and one or more females with their young. The male prevents any of 'his' females from moving too far from him. Kummer (1971) performed the following experiment. Two males, A and B , previously unknown to each other, were placed in a large enclosure. Male A was free to move about the enclosure, but male B was shut in a small cage, from which he could observe A but not interfere. A female, unknown to both males, was then placed in the enclosure. Within 20 minutes male A had persuaded the female to accept his ownership. Male B was then released into the open enclosure. Instead of challenging male A , B avoided any contact, accepting A's ownership.

This result could be explained in two ways. Either B , like a good bourgeois, accepted the rights of ownership, or B could tell that A was the stronger baboon and therefore B did not wish to risk an escalated contest. Kummer was able to show that the former explanation is correct. He separated the two animals. Two weeks later he repeated the experiment with a new female, and with the roles played by A and B reversed. This time B acquired the female, and A did not challenge him. However, escalated contests can take place between male baboons. If a male is removed from a troop, his female(s) will be taken over by another male. If the first male is reintroduced after some days, an escalated fight takes place. Both males regard themselves as the owner of the same female.

Kummer's experiment is an example of a common phenomenon, in which a contest between two animals is settled without escalation. Sometimes the contest is settled without physical contact; sometimes the outcome is determined by a somewhat formalized fight, as when two big-horn sheep collide head-to-head instead of attempting to attack the more vulnerable flanks of their opponent. But it is important to note that fully escalated contests, accompanied by serious injury, also take place.

How are these facts to be explained in Darwinian terms? The strategies which animals adopt are determined by their genes. A strategy which increases an individual's fitness (i.e. which increases its contribution to future generations) will be favoured by natural selection. Difficulties arise because the best strategy to adopt depends on what your opponent will do; it only pays to escalate if your opponent will respond by retreating. In these lectures I shall discuss the kinds of strategy we would expect to evolve. For simplicity, I shall ignore two factors which are relevant in the actual evolution of animal behaviour:

i) Learning - I shall suppose that the two contestants have no previous knowledge of one another;

ii) Kin Selection - I shall suppose that the two contestants are genetically unrelated to one another. This is relevant; for example, an animal which injures its brother reduces its own 'inclusive fitness' by reducing the chances that its brother's genes, which are similar to its own, will be transmitted to future generations.

2. The Concept of an ESS

It is convenient to use some concepts from the theory of games. A 'strategy' is a specification of what an individual will do; it may be a 'pure' strategy (in situation A, always do X) or a 'mixed' strategy (in situation A, do X with probability p and Y with probability q). A 'pay-off' is the change in a contestant's fitness (i.e. expected number of offspring) resulting from a contest. Note that the game theorist's 'utility' is here replaced by 'fitness', a substantial advantage, since changes in fitness can properly be measured on a single linear scale.

The essential concept in what follows is that of an 'evolutionarily stable strategy' or ESS. An ESS is *not* the same as a minimax strategy. Loosely, a strategy is an ESS if a population whose membership adopts that strategy is proof against invasion by any 'mutant' strategy. Clearly, if a population evolves an ESS, it will evolve no further. In this sense we can expect the strategies actually adopted to achieve or be close to an evolutionary equilibrium.

To be more precise, let $E_J(I)$ represent the pay-off to an individual adopting strategy I if his opponent adopts strategy J. Suppose a population consists of a proportion p of individuals adopting I and a proportion q adopting J, where $p + q = 1$. Then if $F(I)$ is the fitness of an individual adopting strategy I, and k is a constant,

$$\begin{aligned} F(I) &= k + p.E_I(I) + q.E_J(I) \\ F(J) &= k + p.E_I(J) + q.E_J(J) \end{aligned} \qquad (1).$$

Now if I is to be an ESS, it is required that, when q is small, $F(I) > F(J)$ for all J. That is, I is an ESS if, for all $J \neq I$,

$$\text{(i)}\quad E_I(I) > E_I(J)$$
or
$$\text{(ii)}\quad E_I(I) = E_I(J) \quad \text{and} \quad E_J(I) > E_J(J) \qquad (2).$$

The conditions in (2) are taken as the definition of an ESS. As will be pointed out in section 4, difficulties arise if a strategy I cannot 'breed true'.

3. The Existence of ESS's

There is nothing in (2) which guaranteed the existence of an ESS for any particular game. In games with a finite number of pure strategies, it is convenient to represent the game in the form of a pay-off matrix. For example, with 3 pure strategies:

		strategy		
		1	2	3
	1	a_{11}	a_{12}	a_{13}
strategy	2	a_{21}	a_{22}	a_{23}
	3	a_{31}	a_{32}	a_{33}

where the pay-off is to the player adopting the strategy on the left. That is, a_{ij} is the pay-off to i played against j. Three conclusions can easily be reached:

i) If any entry on the diagonal, say a_{11}, is greater than all other entries in the same column, then i is a (pure) ESS.

ii) If there are only two pure strategies, an ESS always exists. If there are more than two pure strategies, it is possible to construct a pay-off matrix for which no ESS, pure or mixed, exists.

iii) If I is a mixed ESS, composed of n pure strategies, then the expectation of each of the n strategies, played against I, must be equal. This is a necessary condition for I to be an ESS, but it is not sufficient; I might be an 'equilibrium' strategy, but unstable.

These statements are formulated more precisely, and proved, in a communication from J. Haigh to Maynard Smith (1974).

A game may have no ESS, or it may have more than one; in the latter case, the population will evolve to one or other ESS, depending on its initial composition.

4. Hereditary Mechanisms and the Stability of an ESS.

Consider an infinite population of individuals containing a proportion p_i who adopt strategy i, where $\sum_i p_i = 1$. Let the members of this population play against one another, choosing their opponents at random. The pay-off to an individual adopting strategy I will then be $\sum_j p_j E_j(I)$. If a new generation is constituted by allowing individuals to reproduce proportionately to these pay-offs (or, since the pay-off may be negative, to a constant + the pay-off), we can follow the evolution of the population. We can then ask the following question: if I is an ESS as defined by (2), will a population consisting of a high proportion of individuals playing I be stable in the sense intended by population genetics? This depends on how strategies are inherited, and there are three cases worth considering:

i) Asexual reproduction; individuals can adopt pure or mixed strategies and all strategies breed true. In this case, the conditions specified in (2) are necessary and sufficient to ensure the stability of the population equilibrium. This is the justification for using (2) to define an ESS.

ii) Asexual reproduction; individuals can only adopt pure strategies and pure strategies breed true. Suppose that I is an ESS, and also a mixed strategy; for example, play i with probability x and j with probability $(1-x)$. Then since an individual can play i only, or j only, but not both, a population can realise the ESS only by being a mixture of two kinds of individual, x of i and $(1-x)$ of j.

In this case, if I satisfies (2), it is easy to see that the population is at an equilibrium. Unfortunately, the conditions specified in (2) are neither necessary nor sufficient to ensure the stability of the equilibrium. This awkward fact was pointed out to me by my colleague Dr. C. Strobeck.

iii) Sexual reproduction. Nothing general can be said of this case, except that the actual population will approach the ESS as closely as the hereditary mechanism permits.

In practice, if one suspects the existence of a mixed ESS, one first seeks a strategy I such that $E_I(j) = E_I(I)$ for all the pure components j of I. One can then use (2) to check the stability of I when mixed strategies can breed true, and use computer simulation to check stability when only pure strategies can breed true.

It is an illuminating exercise to investigate the behaviour of simple games. Consider the following, which is based on the children's game of 'rock-scissors-paper'.

		R	S	P
	R	L	2	0
strategy	S	0	L	2
	P	2	0	L

These are the pure strategies. Include also a mixed strategy $V = \frac{1}{3}R + \frac{1}{3}S + \frac{1}{3}P$, which can breed true. If $L < 1$, V is an ESS. But what happens if no V exists? What happens to a population which has, initially, all 4 strategies represented?

5. Some Symmetric Contests

I will now describe a few very simple games which are intended to model some aspects of real contests between animals.

i) 'Hawks' and 'Doves'

Suppose that only two pure strategies are possible:

'Hawk': escalate, and continue to do so until opponent retreats, or until injured.

'Dove': display, but run away if opponent escalates.

Suppose that the pay-off for victory is V , for serious injury is $-D$, and for a prolonged contest is $-T$ to both contestants. Two 'Hawks' are assumed to have equal chances of winning, as have two 'Doves', but the latter each receive $-T$ because the contest is protracted. The pay-off matrix is

	'Dove'	'Hawk'
'Dove'	$\frac{1}{2}V - T$	0
'Hawk'	V	$\frac{1}{2}(V - D)$

If $V > D$, pure 'Hawk' is the only ESS. If $V < D$, let I be the strategy "play 'Dove' with probability p and 'Hawk' with probability $(1-p)$". Then

$$E_I(\text{Dove}) = p(\tfrac{1}{2}V - T) ,$$

$$E_I(\text{Hawk}) = pV + \tfrac{1}{2}(1-p)(V - D) .$$

If I is an ESS, then $E_I(\text{Dove}) = E_I(\text{Hawk})$, i.e. $p = (D - V)/(D + 2T)$. This can easily be shown to be stable. Thus if injury is sufficiently serious, the ESS is a mixed strategy, with 'Dove' becoming increasingly common as injury gets more serious.

ii) 'Retaliator'

Suppose now that a third strategy is possible, which displays unless its opponent excalates, when it escalates in return. The pay-off matrix is

	'Dove'	'Hawk'	'Retaliator'
'Dove'	$\frac{1}{2}V - T$	0	$\frac{1}{2}V - T$
'Hawk'	V	$\frac{1}{2}(V - D)$	$\frac{1}{2}(V - D)$
'Retaliator'	$\frac{1}{2}V - T$	$\frac{1}{2}(V - D)$	$\frac{1}{2}V - T$

If $D > 2T$, the pure strategy 'Retaliator' would be stable against invasion by 'Hawk'; competition between 'Dove' and 'Retaliator' is trivial, since in the absence of 'Hawk' they do not differ. A mixed population evolves towards 'Retaliator'.

The interesting aspect of this example is that retaliation is a common feature of actual behaviour.

iii) The War of Attrition

Suppose now that only display is possible. The winner is the animal which persists the longer. A long contest costs both contestants a negative pay-off, which increases with the length of the contest. The only choice of strategy open to an animal is the duration of its display. Suppose that two contestants A and B choose to continue for times T_A and T_B, where $T_A > T_B$. To each of these times there corresponds a pay-off, $-m_A$ and $-m_B$, which is the cost of a contest of that length. V is the pay-off for winning.

Then the pay-off to A is $V - m_B$, and to B is $-m_B$. Note that m_A is irrelevant so long as it is greater than m_B.

Is there an ESS? Clearly no pure strategy can be an ESS (one assumes that if $m_A = m_B$, then each player receives $\frac{1}{2}V - m_B)$. It is possible to find a stable mixed ESS. Thus if the ESS is to choose m between x and $x + \delta x$ with probability $p(x)\delta x$, then

$$p(x) = \frac{1}{V} e^{-x/V} . \qquad (3)$$

It is common to find that in symmetric games such as those discussed above, the ESS is a mixed one, containing a mixture of meek and bloody-minded strategies.

6. Asymmetric Contests

In most actual contests between animals, there will be a difference in size or weapons between the contestants, or in the pay-off for victory; for example, in a contest over food, one animal may be hungrier than the other. If neither contestant is aware of this asymmetry, it cannot affect their choice of strategy, although, of course, a difference in size may affect the outcome. Usually one or both contestants will have some information about the asymmetry. In this case the ESS will be altered.

For simplicity, consider a contest with an 'uncorrelated' asymmetry; that is, one which is not associated with a difference in pay-off or fighting ability. Take, for example, the baboon contest described in §1. The contest was associated with an asymmetry between 'owner' and 'late-comer', but this was not correlated with any difference in pay-off to the winner (the female would be equally valuable to A or B), or in fighting ability (at least, if there was such a difference, the animals behaved as if they knew nothing about it).

At first sight it might seem that an uncorrelated asymmetry could not affect the choice of strategy, but this would be mistaken. Let us consider the War of Attrition game, but assume that there is a 'label' associated with each contest, so that A 'sees' the label 'Red' and B 'sees' the label 'Blue'. A particular individual is equally likely to see Red or Blue, but knows that his opponent is seeing the opposite colour. The label is always present, but has no effect on the rules of the game.

Let J be the strategy defined by (3). Let I be the strategy

'If see Red, play M, where $M > V$;

if see Blue, play 0.'

Then it is easy to show that I is stable against invasion by J, but that J is not stable against I. In fact, I is now the ESS. There is also a second ESS, I', which plays M if sees Blue and 0 if sees Red. A population can evolve to I or to I'.

The most obvious application of this example is to the asymmetry between 'owner' and 'late-comer', as in the baboon experiment.

7. Games of Imperfect Information

The asymmetric game considered in the last section has a simple solution because both contestants have complete information. An animal which sees Blue knows that its opponent is seeing Red, and vice versa. The 'Hawks' and 'Doves' game would also have a simple ESS if each contest was associated with a difference in size which was clearly and correctly perceived by both contestants. The ESS would be 'escalate if you are larger; retreat if you are smaller'; this would be stable even if size was a poor guide to the outcome of an escalated contest.

Things become much more difficult if the information available to the contestants about the asymmetry is incomplete or unreliable. Thus it does not matter if the size difference is a poor guide to the outcome of escalation, but it is crucial if an animal's estimate of the size difference is itself faulty, because there is then the possibility that each contestant may estimate that it is larger than its opponent. I have made a start at analyzing such contests, and also in analyzing contests in which information is acquired only during the course of the contest. Further, as soon as asymmetries enter into the settling of contests, the possibility of 'bluff' arises. If an animal backs down because it estimates that an opponent is the larger contestant, then it may pay an animal to appear larger than it really is.

Finally, to illustrate how difficult such games can be, consider the following simple extension of the War of Attrition game. Suppose the prize is an item of food, and the pay-off depends on how hungry you are. You 'know' how hungry you are yourself; that is, you know that the pay-off to you for victory would be V, say; V will vary from contest to contest. You do not know how hungry your opponent is, but you do know the 'distribution of hunger'; that is, you can assume that your opponent has a pay-off, V', which is drawn randomly from the distribution with density function $p(v)$, where $\int_0^\infty p(v)dv = 1$. As it happens, V is also drawn from the same distribution. How should you choose your strategy? More precisely, what choice of strategy is an ESS? Common sense suggests that you should play a higher value of m when V is higher than usual, but how much higher? I have been unable to make progress with this problem, even by selecting a distribution $p(v)$ for mathematical convenience.

ECOLOGICAL MODELS IN EVOLUTIONARY TIME

One obvious approach to the modelling of an ecosystem is to take as the state variables the 'densities' (i.e. numbers per unit area) $N_1, N_2, \ldots, N_m$ of the m species composing the system. The dynamics of the system can then be described by the equations

$$dN_j/dt = F_j(N_1, \ldots, N_m) \quad , \quad (j=1,\ldots,m) \quad . \quad (1)$$

This approach was pioneered by Lotka and Volterra, and has recently been developed by Kerner, May and others. Apart from the disadvantage that one cannot in general solve sets of equations of this kind, the description suffers from various other drawbacks. In particular:

i) The environment is treated as constant in time;

ii) No allowance is made for spatial extension, let alone for spatial heterogeneity of the environment;

iii) In describing the density of a species by a single variable N_i, no allowance is made for differences of sex, age, or genotype. Of these, the last is perhaps the most important; by ignoring age structure, one omits the time delays which are probably the major cause of the oscillations of real ecosystems.

It is, of course, unhelpful to complain of a set of equations which are too complicated to solve that they are not complicated enough. I do so because it is important that mathematicians should understand at the outset that these equations may already be too abstract to be of much use in many problems.

One awkward conclusion which emerges from a study of equations (1) is that they are unlikely to describe a stable system. This problem has been discussed in a number of recent papers, and I will not go into it again. The essential conclusion, however, appears to be that an arbitrary set of equations is unlikely to lead to a stable equilibrium if m is large.

An ecologist would reply that the interactions between species are not arbitrary, but are the outcome of millions of years of natural selection. This reply is correct, but does not immediately resolve the difficulty. Natural selection does not act by favouring stable ecosystems at the expense of unstable ones, and only rarely by favouring 'prudent' species at the expense of imprudent ones. In general, selection acts to favour individuals of high 'fitness', that is, with a high probability of survival and successful reproduction. Thus if some genetic change increased the individual fitness of rats, for example, this change would spread through the population, even if the long term result was the destabilization of the ecosystem of which rats form a part.

We have to recognize that the constants which appear in ecological equations are constants only in short-term, ecological time. In long-term, evolutionary time they are variables. Thus, in Lotka's equations for two-species competition,

$$dx/dt = x(a - bx - cy)$$
$$dy/dt = y(e - fx - gy) \quad , \qquad (2)$$

the constants a,b,c,e,f,g represent characteristics of competing species which will change in evolutionary time. This raises the following problem. Given that these constants change, and under natural selection tend to maximize the fitness of individuals, what effect will this have on the stability of the system as a whole? Can we account for the stability of ecosystems in terms of individual selection? Some ecologists (e.g. Pimentel, 1968) have argued that we can; others (e.g. Wynne-Edwards, 1962) have argued that we cannot, and have invoked other selective mechanisms ('group selection') to get us out of our difficulty. As yet, little progress has been made with equations such as (1) when the functions F_j are subject to evolutionary change.

The paper by Lawlor and Maynard Smith (1976) is one such attempt. It concerns the coevolution of two 'consumer' species competing for two renewable resources. The detailed arguments of that paper will not be repeated here, but it may be helpful to comment on the 'maximization principle' which has been used.

Suppose that N_i is the density of the ith species in an ecosystem, and that

$$dN_i/dt = N_i\phi_i(a_{i1},a_{i2}\,;N_1,\ldots,N_m) \quad (i=1,\ldots,m), \qquad (3)$$

where a_{i1},a_{i2} are two ecological parameters which can be treated as constants in ecological time, but which change in evolutionary time as a result of changes in the genetic constitution of species i.

Now a_{i1} and a_{i2} are interdependent, and subject to some constraints. In the simplest possible example, a_{i1} and a_{i2} represent the proportions of time which species i spends looking for the only two available resources; hence $a_{i1} + a_{i2} = 1$ for all i. More generally, we could suppose that species i is restricted to some area on a plane. (There is no reason why only two such interdependent parameters should exist, but we have not attempted to extend our analysis to n dimensions.) Levins has called such an area a 'fitness set'; it is the set of phenotypes which the species can achieve.

To what point within this area will the species evolve? Instead of seeking a trajectory, we shall be satisfied to find a stable equilibrium point, i.e. an 'evolutionarily stable strategy' or ESS. The concept of a strategy is appropriate because the optimal policy for one individual depends on what others are doing; if everyone else is eating cabbage it pays to eat spinach, and vice versa.

We therefore assume that the ESS will lie at some point, say (a^*_{i1}, a^*_{i2}), on the line bounding the fitness set, (see Figure 1).

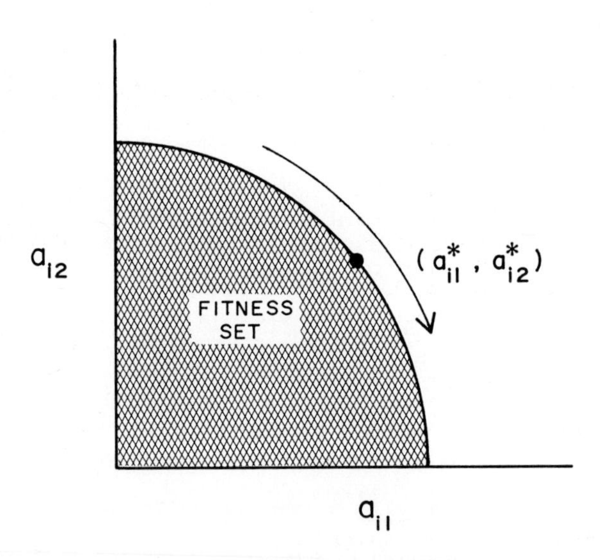

Figure 1

Imagine that a population has evolved to that stable point, and that the other species in the ecosystem have also evolved to their ESS's. Suppose also that the species densities have achieved the corresponding ecological equilibrium values, $N_1^*, N_2^*, \ldots, N_m^*$, determined by the equations

$$\phi_i[a_{i1}^*, a_{i2}^*; N_i^*, \ldots, N_m^*] = 0 \qquad (i=1,\ldots,m) \quad .$$

Then, since the equilibrium point is an ESS, individuals with the phenotype (a_{i1}^*, a_{i2}^*) are the fittest possible. In other words, if we consider a point moving along the boundary of the fitness set (consider the arrow in Figure 1), the fitness of the corresponding phenotype will be maximized as it passes through the ESS. That is,

$$\left[\frac{\partial \phi_i}{\partial a_{i1}}\right]^* = 0 \quad ,$$

where the differential is evaluated at the ESS; clearly, difficulties arise if the ESS lies at a cusp or corner.

In the case of two competing species, we have compared the ESS's adopted by each consumer species in isolation, with the ESS's adopted when both are present together. It turns out that coevolution does lead to greater resource specialisation, and hence to greater stability. However, it would be dangerous to generalize this conclusion. Difficulties arise at once if we consider the coevolution of a predator and its prey. Volterra's equations for this are

$$dx/dt = ax - bx^2 - cxy$$

$$dy/dt = -ey + cxy \quad , \qquad (4)$$

where x and y are the respective prey and predator densities (scaled to ensure that the consumption of one prey is sufficient to make one predator equivalent).

Continued coexistence in ecological time requires that c, which measures the predator's prey-catching efficiency, remains within certain limits. Thus if $c < eb/a$, the predator will die out. If c is too large, the prey density will decrease. In a finite stochastic model, this change might lead to extinction; in more realistic formulations, it is likely to result in large amplitude fluctuations and an enhanced risk of extinction.

These considerations prompt the question "Are there evolutionary processes which will tend to increase c when it approaches eb/a, and which will reduce c when it becomes too large?".

Now c can be altered by genetic changes either in the predator or the prey. The influence of selection on the prey will tend to reduce c, whereas predator selection should increase prey-catching efficiency. Which will win?

Rosenzweig (1973) has suggested a very simple reason why c might approach eb/a and then begin to increase. Initially, few prey are being killed; therefore, the influence of selection on the prey will not be very intense. However, predator selection will be intensive, hence c will increase. If c should become too large, Slobodkin (1974) has suggested the following process. No predator relies on exactly one kind of prey. A more realistic model would be

$$dx_1/dt = a_1x_1 - b_1x_1^2 - c_1x_1y$$
$$dx_2/dt = a_2x_2 - b_2x_2^2 - c_2x_2y \qquad (5)$$
$$dy/dt = -ey + c_1x_1y + c_2x_2y \quad .$$

The predator is handicapped by the need to increase both c_1 and c_2, whereas each prey species is concerned with only one of c_1, c_2. If the prey differ sufficiently, so that predator adaptation which tends to increase c_1 will decrease c_2, and vice versa, the prey may be safe.

The argument, although not formal, is persuasive. But suppose there are several predators and several prey? Antelopes are eaten by cheetahs, which outsprint them, and by hunting dogs, which outlast them. Clearly, real ecosystems are complex, and in order to understand the evolution of their component species it may be necessary to take notice of this natural complexity. In formal terms, this means tackling the problem which Lawlor and I consider in a complex system instead of a two-species system. Even if the mathematical difficulties could be overcome, we would first have to find a plausible set of constraints.

At present I see no way round these difficulties. We still have much to learn by modelling simple ecosystems with few species. The multi-species problem is more formidable. My present hunch is that we shall also have to account for the hierarchical structure of such systems. Our models may have to be formulated in terms of different state variables - for example, the biomass of herbivores or of detritus feeders. What place will evolutionary concepts take in such models? I suspect that we shall have to appeal to natural selection for the answers to many problems; for example, why only certain categories of herbivores exist may be explained in terms of their ecological efficiencies, intrinsic rates of growth etc. These remarks are very vague; obviously I would be more precise if I could.

REFERENCES

Kummer, H. (1971) *Primate Societies*. Chicago: Aldine-Atherton, Inc.

Lawlor, L.R. and Maynard Smith, J. (1976) The coevolution and stability of competing species. *American Naturalist, 110,* 79-99.

Maynard Smith, J. (1974) The theory of games and the evolution of animal conflicts. *Journal of Theoretical Biology, 47,* 209-221.

Pimentel, D. (1968) Population regulation and genetic feedback. *Science, 159,* 1432-1437.

Rosenzweig, M.L. (1973) Evolution of the predator isocline. *Evolution, 27,* 84-94.

Slobodkin, L.B. (1974) Prudent predation does not require group selection. *American Naturalist, 108,* 665-678.

Wynne-Edwards, V.C. (1962) *Animal Dispersion in Relation to Social Behaviour.* Edinburgh: Oliver and Boyd.

The Generation and Recognition of

Patterns in Biological Systems

Professor R. Rosen

Killam Professor of Biomathematics
Department of Physiology and Biophysics
Faculty of Medicine
Dalhousie University
Halifax, Nova Scotia

THE GENERATION AND RECOGNITION OF PATTERNS IN BIOLOGICAL SYSTEMS

General Introduction

Form and pattern are probably the most obvious, immediate and pervasive characteristics of organisms, and indeed, of all aspects of the world around us. Problems involving the description, generation, and maintenance of form have been studied continually, and with ever more powerful and subtle techniques, for literally thousands of years. Yet today, despite all this, and despite all the data which has been collected, I believe it is fair to say that the forces responsible for the production and the stabilization of biological form remain obscure. To be sure, we can easily observe, and marvel, at the effects of these morphogenetic forces wherever we look in the biosphere. But to characterize these forces in quantitative terms, to isolate their properties and predict their effects, has proved to be quite a different matter. Many of the great controversies of biology, such as the mechanist-vitalist or the preformation-epigenesis controversies, owe their origin primarily to disagreements over the proper way to study form. And even today, there is no real consensus on how the fundamental problems posed by form and its generation are best approached.

The present lectures represent an attempt to survey the chaotic but fundamental areas of form and pattern generation and recognition, in such a way that they become accessible to those interested in the relation of mathematics to the life sciences. We believe that no resolution, or even proper formulation, of these problems can be forthcoming, unless the mathematician and the physicist can bring their skills, techniques and insights to bear upon them. For, as D'Arcy Thompson, one of the pioneers in this area, pointed out in his classical book, "*On Growth and Form*", in 1917: "Cell and tissue, shell and bone, leaf and flower...Their problems of form are in the first instance mathematical problems, their problems of growth are essentially physical problems, and the morphologist is, *ipso facto,* a student of physical science". These words are, if anything, even more true today than when they were first enunciated.

The plan of these lectures is as follows. In the first lecture, we shall attempt, so to speak, to get an idea of what we are up against. Some selected examples of form and its generation are drawn from all areas of biology, and the experimental facts regarding them described. We attempt to draw some commonalities from these examples, diverse as they are, and then briefly consider one class of attempts to explain such phenomena. These attempts ultimately devolve on evolutionary considerations, regarding form as generated by natural selection; thus on the one hand, organic form becomes subject to considerations of functional optimality, and on the other hand, related forms can be regarded as deformations or models of one another.

In Lecture II, we turn from the empirics of biological form and its generation to a review of the various mathematical approaches which have been made, mainly over the past 15 years, to morphogenesis. These approaches primarily view the generation of form as arising from interactions between subunits and substructures. We can view form and pattern as arising (a) from movements of these units with respect to each other; (b) from differential changes in the properties of the units; and (c) from differential changes in the relative numbers of these units. Each of these three basic morphogenetic mechanisms, which were exemplified on many levels in the systems considered in Lecture I, can be described in formal mathematical terms, and the capacities of each of them to generate pattern and form can begin to be assessed. We also touch briefly here on the recent work of Thom, which is of a rather different character.

Lecture III turns back towards biology, with a consideration of the genome and its role in biological systems. As we indicate, present views of primary genetic activity are mainly biochemical. We briefly review the currently accepted views of genetic activity, and show how these can be treated in mathematical terms. We see that there is an enormous chasm between the biochemical aspects, directly under genetic control, and the morphogenetic aspects we have described earlier. We initially pose the crucial problem: how are we to integrate the genome into the control of morphogenesis?

In Lecture IV, we begin to approach this problem with a sketch of epigenetic mechanisms. These may be regarded as mechanisms arising from the interactions of gene products. All of the morphogenetic mechanisms described in Lecture II are epigenetic mechanisms; the role of the genome is thus, ultimately, to set initial conditions and constraints for the dynamics which govern epigenetic phenomena. This observation, in turn, raises some profound conceptual questions, particularly bearing on the relation between the theoretical treatment of morphogenesis and experimental observation.

Lecture V contains a development of a possible integrated theory of genome-directed morphogenesis. The point of departure here is the hypothesis that one cannot generate a pattern in a system without, in some sense, recognizing a pattern in an associated system, and conversely. We thus attempt to view the generation of global morphogenetic patterns as arising from the recognition of an underlying genetic pattern. In order to do this, we develop some of the considerations underlying the recognition, or measurement, or classification, or discrimination, of classes of patterns. Here too, profound conceptual matters, bearing on the relation of systems to their subsystems, and on the empirical interpretation of mathematical formalisms, must be considered. But we hope to present at least a first approximation to what a genome-directed theory of morphogenesis must be like.

The reader will note that these lectures stress, perhaps more than other treatments of morphogenesis, the conceptual underpinnings of mathematical representations rather than their technical development. This is done from a conviction that the really difficult problems of science, and of biology in particular, are conceptual problems, many of which devolve ultimately upon the relation of an abstract representation of a system to the real system it describes, and upon the relation of several such representations to each other. Until these basic epistemological questions are recognized, we will have no firm basis for proceeding. The mathematician, in particular, is accustomed to seeking what can be discovered within the confines of a particular formalism. The theoretical scientist, who is concerned precisely with creating such formalisms, must also seek to understand what properties of his systems *cannot* follow from a specific formalism, and how separate formalisms, which each embody some aspects of reality of his systems, can be integrated to obtain a more complete picture. The area of morphogenesis illustrates all of these aspects in a particularly pregnant way, and it is with these aspects that we have attempted to come to grips, both in the context of the problems we really wish to solve, and of the formalisms which have been advanced over the years as partial solutions.

Lecture I

As noted in the Introduction, problems of form are the most obvious and pervasive features of biological systems. When I say this, I am sure that my readers will have a reasonably clear intuitive idea of what I mean. Yet, as with almost all of the important features of biological systems, if I were required to *define* form in a concrete, quantitative way, I would be at a loss. Consequently, to be as certain as possible that we all begin with the same idea as to our subject-matter, and at the same time to point out the diversity and generality of that subject-matter, it is best to begin with a series of specific biological examples of form and form-generation.

Form at the Molecular Level

We shall begin by considering form and pattern at the most microscopic level. Already here, as we shall see, notions of form play the most fundamental role. Of the many illustrations we could give, we shall consider here those involved in the study of the *conformations of proteins*.

Proteins are the workhorses of biology. They constitute the specific catalysts which determine which chemical reactions shall occur in an organism, and how fast they shall proceed. They seem to be involved crucially in all molecular processes in which *recognition* and *specificity* are manifested, such as enzyme-substrate, antigen-antibody, or drug-receptor phenomena. In addition to this, they play a variety of structural roles. Collagen, the universal supporting matrix for cells in multicellular organisms, is protein, as are such organelles as flagella and cilia. The structural basis for the biological activity of muscle lies in its proteins.

The simplest protein molecules constitute a family of linear polymers, or polypeptides, built from a family of about twenty biologically important monomeric units, or amino acids. The order in which the amino acids are arranged in the molecule is called the *primary structure* of the molecule. As we shall see, it is this primary structure which is specified by the genome. A polypeptide is synthesized, amino acid by amino acid, starting at one end and ending at the other.

A nascent polypeptide appears in a biological system as a random coil. In this form, the molecule has no biological activity. All of its specific activities, and the capability to carry out the various functions we have mentioned, are only exhibited when this randomly coiled polypeptide assumes its ultimate *conformation,* its specific, active three-dimensional form. It assumes this active conformation through a morphogenetic process of *folding*. In the folded structure, there is no obvious symmetry or any other simplifying feature. Moreover, it is not at all apparent how, from the geometry of such a molecule, one can infer anything about its biological activities (e.g. its "active sites").

The morphogenetic process of folding seems to be a spontaneous one. Thus, for instance, we can experimentally re-convert a folded polypeptide back into a random coil, e.g. by raising the temperature; this process is called *denaturation*. On allowing the system to cool back to ordinary temperature, it is found that the biological activity is fully restored; the polypeptide has re-folded back to its active conformation. Since this process takes place spontaneously, it represents an instance of *self-organization*. The folding is quite rapid; it proceeds to completion in minutes (Anfinsen, 1968).

It is a dogma of molecular biology that the conformation of a polypeptide (often called its *tertiary structure*) is completely determined in a given environment by its primary structure. On the basis of the physical and chemical properties of the amino acids themselves, numerous authors, especially Scheraga (1969) have attempted to calculate this tertiary structure, on the basis of the assumption that it represents a minimal free-energy state of the molecule. However, simply to write down the potential function for such a molecule, even in approximation, is discouragingly difficult; to then proceed to minimize it is perhaps beyond current capabilities.

Although primary structure may be assumed to uniquely determine tertiary structure, the converse is not quite true. Corresponding proteins taken from different organisms are often substantially different from each other in primary structure, yet their tertiary structures, and biological activities, are at least very similar. Moreover, in the denaturation experiments mentioned above, it is possible to excise long pieces from the polypeptides involved, thereby drastically changing the primary structure, and yet the mutilated polypeptides re-fold in such a way as to recover most of their biological activity (Anfinsen 1968).

Thus, already at the most microscopic level, we see that problems of form and morphogenesis lie at the root of everything else. The ideas we have outlined regarding polypeptides suggest: (a) that morphogenesis is a process governed by the interactive capabilities of subunits; (b) that polypeptide conformations can be quite *stable* to various kinds of perturbations, either of ambient conditions like temperature, or to the dynamical properties of the morphogenesis which gives rise to active conformations in the first place.

Although many biologically active proteins are single polypeptides, many others are aggregates of several polypeptides. Hemoglobin, for instance, consists of four polypeptides (two pairs of like chains); insulin consists of two distinct polypeptides. Each constituent polypeptide of such an aggregate is synthesized separately; their association is determined by the tertiary structures of the constitutent chains, and, like the folding of an individual chain, represents a process of self-organization. Experiments such as those described for polypeptides, such as denaturation and renaturation, modifications or amputations of primary structures, and hybridization of chains from different sources, may be performed, with results similar to those we have already mentioned. Thus, the morphogenetic processes which give rise to these assemblies also exhibit remarkable stability properties.

Let us consider some other associations of polypeptides, which are important in generating morphology at higher levels.

(a) Microtubules and Microfilaments

Filamentous processes are found everywhere in biology. At the lowest level, we find ultramicroscopic tubular and filamentous structures ubiquitous in cells, and playing fundamental roles in the most basic cellular processes. Microtubules, for instance, form an essential constituent of the mitotic spindle, and ultimately are responsible for the correct movement of the chromosomes at cell division. The same tubules are ubiquitous constituents of cilia and flagella, the organs of cellular locomotion. They are also to be found in the processes relating to cellular form.

The basic constitutent of microtubules, wherever they may be found, is a protein (tubulin), which is itself apparently a dimer consisting of two polypeptide chains. The assembly and disassembly of microtubules depends on the specific aggregation and de-aggregation of tubulin; this can take place spontaneously, but also seems to be under careful control through obscure cellular mechanisms. The behavior of microfilaments is similar, though here the basic subunit is closely related to the protein *actin*, one of the essential constituents of muscle in higher organisms, (and which, together with the protein myosin, is responsible for its motive properties). The reader interested in further details regarding microfilaments and microtubules can consult Spooner (1975); for muscle, the reader may consult Huxley (1967).

(b) Viruses

A virus is essentially a core of genetic material (DNA or RNA) surrounded by a coat of protein. This coat apparently serves a protective function, as well as allowing the virus to recognize its host cells. The simplest viruses are the "spherical" viruses, consisting of a polyhedral coat surrounding the genetic core. This coat in turn consists of identical protein subunits packed in an orderly way to generate the coat. Considerations of symmetry (Crick and Watson,1956; Klug,1969) place limitations on how these units can pack (this represents one of the few areas in which symmetry considerations have played a direct role in biology). The formation of these viruses seems once again to be a self-assembly; the individual subunits are separately synthesized and spontaneously aggregate. If the final structure is artificially separated (randomized) under physiological conditions, the subunits will re-aggregate to form virus particles structurally and functionally indistinguishable from the native virus.

Another virus, exhibiting helical symmetry, is the familiar tobacco mosaic virus. Here again, a helical core of nucleic acid is surrounded by a protein coat consisting of identical subunits. The virus will re-assemble spontaneously if it is disaggregated. The coat protein alone, in the absence of RNA, will re-aggregate to form structures indistinguishable from native virus, except that the length of the structures is variable. (Caspar and Klug, 1962)

A much more complex virus is the T4 bacteriophage. This complex structure involves a large number of different proteins, exhibiting correspondingly different functions. Unlike the others described so far, the bacteriophage is not produced through a mechanism of self-assembly, and if disaggregated, it will not re-form (although certain substructures do apparently possess this property). Since the genetics of this virus, and of the host cell (E. coli) have been developed in great detail, and since the morphogenesis of the virus depends directly on molecular properties coded directly by the genome, the bacteriophage provides a tantalizing means of approach to the genetic control of form and its generation (cf. Levine, 1971).

(c) Cellular organelles

We have already mentioned the locomotory organelles of cilia and flagella. Many other important organelles, such as the mitochondria, the chloroplasts and the ribosomes appear to be analyzable into substructures, and capable of at least a limited capability to re-assemble after disaggregation (cf. Warren, 1967). These organelles contain many other kinds of molecules besides proteins (e.g. lipids for their membranes). We cannot go further into the genesis of these crucial little structures here, but it is useful to note their complex structure and organization, even in such a microscopic realm.

Form at the Cellular Level

The well-known cell theory asserts that the single cell is the basic unit of biological structure and organization. Cells exist in a bewildering variety of forms, ranging from the bacteria, through free-living and parasitic protozoa, algae and fungi, to the varieties of cells found in multicellular organisms.

The cell does not possess the remarkable properties of self-organization which we have seen at the molecular level; it is not presently possible to dissociate a cell into parts which will spontaneously re-assemble to reconstitute the cell. However, it is possible to perform many other experiments similar in character to those we have already seen. It is possible to exchange materials between cells of different origins; to replace one nucleus by another; one cytoplasm by another; one membrane by another. It is possible, using microbeams and microsurgery, to ablate or kill specific regions of cells. It is possible to fuse two cells, to create one viable cell incorporating all the parts of its constituents. From such experiments, other remarkable stability properties emerge.

A classic cell on which experiments of this type can be carried out is the ciliated protozoan, *Stentor*. This single cell is highly organized, with a characteristic morphology. It turns out that this morphology is exceedingly difficult to disrupt by any of the experimental approaches just mentioned. The cell will regenerate its characteristic form following surgical ablation, following the grafting of fragments from itself or other cells into different configurations, or following the fusion of cells (cf. Tartar, 1961). The same is true of other ciliated protozoa, such as *Paramecium* and *Blepharisma*.

Another cell, vastly different from Stentor, is the one-celled alga *Acetabularia*. This is an enormous cell, shaped like a mushroom, which grows to the length of centimeters. It contains a single nucleus, located in the root area. If the cap of the cell is amputated, a new cap will form, identical to the original. If one attempts to graft a new cap, from a different species, onto the stalk, that cap will be resorbed, and a new cap, characteristic of the original cell, will be generated (Hammerling, 1946). Here again, the form has remarkable regulatory properties, which appear ultimately to devolve upon the nucleus, and hence upon the genome.

Cells and Multicellular Aggregates

Multicellular aggregates, even of free-living cells, often appear to exhibit a characteristic form. A bacteriologist can often identify a species of bacterium by looking at the kind of colony it forms on a nutrient medium, without microscopic investigation. Algae often form multicellular aggregates of particular types; from filamentous forms such as *Spirogyra,* to the famous Volvocine series, consisting of spherical aggregates of different sizes. The most complex,*Volvox,* has most of the properties of an independent multicellular organism, though it is usually regarded primarily as a colony.

Associations may occur between cells of different types, associated in symbiosis, just as an oligomeric protein may consist of polypeptide chains of distinct species. Characteristic associations of algal and fungal cells, called *lichens*, are well known

A remarkable case of cell association is that presented by the slime-mold *Dyctiostelium.* In environments containing an excess of food (bacteria), the organism consists of a multitude of individual free-living cells resembling amoebae, uniformly distributed over the medium. When food becomes scarce, these cells spontaneously aggregate through a chemotactic process, to form clumps, or slugs; these multicellular aggregates then migrate as a unit. Ultimately this initially homogenous mass differentiates to produce a stalk containing spores, which then scatter widely. Each produces a single amoeba, and the process repeats. (This organism has been extensively investigated; cf. Bonner, 1967).

Cells and Developmental Phenomena

Striking as the previous considerations have been, they are eclipsed by the morphogenetic processes occurring in *development,* that process by which a single cell (a zygote, or fertilized egg) gives rise to an entire multicellular organism. Naturally, we cannot review the subject of development in a few pages; we shall instead merely cite some experimental observations on developing systems which are typical. For fuller details, we can refer to any number of excellent references, such as Child (1941), Dalcq (1938), Waddington (1954, 1962), Weiss (1939, 1968).

At the level of the fertilized egg, quite a number of experiments can be done. It was early found that there seem to be two different classes of fertilized egg. One class, called *mosaic eggs*, seems very highly organized; any attempt to interfere with the structure of the egg (e.g. by amputation of a region of the egg) will result in the production of a defective or incomplete organism. The other class of eggs, called *regulation eggs,* does not respond in this way; if a fragment of the egg is removed, or if its contents are randomized by centrifugation, a complete and normal organism will still result.

At the blastula level, similar results are found. In general, if an intact blastula is left to develop, its specific cells will give rise to descendants of precisely determined biological characteristics; such "fate maps" may be found in any embryology text. Thus, one cell in the aggregate will give rise to cells entering into such structures as skin and its associated organs; another will give rise to bone and muscle; another to nerve; another to visceral organs, etc. In a blastula arising from a mosaic egg, any interference with the spatial order of the cells in the blastula will produce a defective organism; on the other hand, in a blastula arising from a regulation egg, randomization of these cells (i.e. modifying their initial spatial ordering) will cause the different cells, which would originally have developed along one path, to switch over to a different path, in such a way that a normal organism is produced. Still further, the blastomeres of a regulation egg can be separated from each other; each will still give rise to a normal (though smaller) embryo. Such regulatory phenomena seemed so far removed from any physical explanation that the great embryologist Driesch was led to postulate a vitalistic factor, or entelechy, to account for them. Perhaps the most spectacular experiments in this connection are, however, those involving the fusion of blastulae arising from different parents, as in the allophenic mice of Mintz (1967); here too, development will proceed normally, and give rise to an organism which is a genetic mosaic. Such mosaic organisms are currently being much studied to obtain insight into the genetic control of development.

Subsequent to gastrulation, much of this plasticity is lost. A dissociated gastrula will not re-aggregate to produce a viable developing system. However, if a fragment of a gastrula, which is already differentiated into two cell types (ectoderm and endoderm) is removed from the embryo and its cells randomized, it is found that the cells will reorganize themselves into a blastula-like structure, in which all the ectodermal cells are on the outside of the aggregate, and all the endodermal cells are on the inside. This, of course, is the arrangement present in the original developing system. Moreover, if at a later stage of development, organs such as lung or kidney, which possess elaborate histologies and are composed of many cell types, are removed and randomized, it is found that the cells will re-aggregate in such a way as to re-establish a recognizable approximation to the histology of the original organ. Such behavior is called *sorting-out,* and has been studied under many conditions (cf. Trinkaus,1969; Curtis, 1967). The sorting-out behavior is also exhibited in simple multicellular systems; it was shown by Wilson in 1907 that sponges, the simplest multicellular organisms, could be disaggregated into their constituent cells; these cells would then spontaneously reassemble to form small viable sponges.

To test the loss of "totipotence" in the cells of developing systems, Briggs and King (1959) removed the nucleus from cells of blastulae and gastrulae, and returned them to the cytoplasm of an enucleated fertilized egg. It was found that the capacity of such a nucleus to induce normal development progressively diminished with "older" nuclei (but significantly, there was always some percentage of normally developing eggs obtained from such nuclei).

It was discovered in the 1920's, by Spemann and others, that normal development depended in many cases upon the interaction of two regions of the embryo. This interaction involves a region called an *organizer*, which produces an inductive or evocative stimulus, and a region *competent* to receive this stimulus, and respond to it by differentiating in an appropriate way. This discovery, and its relation to chemical gradients, could be tested by numerous transplantation experiments, in which regions were removed from one part of an embryo and grafted onto other regions, or regions removed from an embryo at one stage of development could be transplanted to corresponding or different regions on an embryo at a different stage. The results of such experiments gave rise to the various *field theories* of development, which are still actively pursued.

Closely allied with developmental phenomena are those involving *regeneration*. This term refers to the response of an organism to the amputation of some part or organ. We have already seen phenomena of regeneration at lower levels of organization; the renaturation of protein is a form of regeneration, as is the response of *Stentor* or *Acetabularia* to amputation of anucleate fragments. The same phenomenon persists in higher organisms. Such animal forms as *Hydra* are well known for their regenerative capacities, and indeed are widely studied (cf. Wolpert, 1969) for the light that these capacities throw on morphogenesis. The planarian flatworms are also well known for the capacity of even quite small fragments to give rise to intact, fully differentiated organisms. Among plants, this capability is manifested even more strikingly; a single somatic cell from a higher plant can, under appropriate conditions, give rise to an entire plant. Among higher animals, however, the regenerative capabilities tend to become more limited; the salamander is the highest organism that can ordinarily regenerate an amputated limb (cf. however Singer, 1968), while man can regenerate only so far as to heal relatively superficial wounds. However, it is clear that there is a close relationship between regeneration and morphogenesis (cf. Faber, 1971).

Extracellular and Extra-Organismal Patterns

Many single-celled organisms have the capacity to generate extracellular patterns in inorganic material i.e. to form shells of silica (as in the diatoms) or calcium carbonate (as in the foraminifera). These mineral shells are produced in species-specific, often exceedingly intricate forms, primarily for purposes of support and protection.

The formation of such extracellular support structures, in correspondingly species-specific forms, is common among multicellular organisms also. For instance, the forms produced by colonial coelenterates known as corals are well known. Likewise, the external shells of mollusks (clams , snails, etc.), brachiopods, and cephalopods such as *Nautilus* are familiar to all. There is not much difference between such external skeletons, produced by individual organisms and by colonies, and the nests or hives built by colonial insects or higher forms. And functionally, such external supporting structures are analogous to the external skeletons of insects, or the internal skeletons of vertebrates. Some of the most closely studied patterns in biology involve such species-specific skeletal structures, such as the spacing of bristles on the external skeletons of an insect (Stern 1956; Wigglesworth,1945; Locke,1967). These spacing patterns, in turn, are formally similar to those arising in the spacing of stomata on leaves, or the spacing of trees in a forest, or, at another level, the spacing of cities in a landscape.

The above, necessarily superficial review of the biological aspects of form and pattern has hopefully served to indicate the breadth of the problems involved. We have seen that these extend from the conformation of a protein molecule to the properties of a coral reef. It might be thought that there could be nothing in common between such diverse systems but we hope to indicate that this is not so. Indeed, the few examples we discussed were chosen in such a way as to indicate a number of important commonalities which will be the bases for most of our future discussions. Let us extract some of these commonalities now; their more precise formulations, and their consequences, will be taken up in the subsequent lectures.

(a) In all cases we have discussed, the terms "form" and "pattern" have referred to the disposition of subunits, or sub-structures, with respect to one another. The processes of form and pattern generation, or morphogenesis, refer to dynamical processes arising from the internal properties of these subunits, and from their interactions with one another. As we shall see abundantly, especially in Lectures IV and V, to analytically decompose complex structures and processes into simpler subunits, and then to reconstruct the properties of the original system from those of the subunits, is a general conceptual strategy, applicable to any kind of process. We obtain ample motivation for pursuing such a strategy from the examples we have considered.

(b) The experimental approaches to morphology and morphogenesis which were described above also follow a definite pattern, at whatever level of organization we may consider. In general, given any unknown dynamical process, the only way we have of studying it, apart from simple observation, is to perturb the process and observe the transients. In the situations we have described, in which form and morphogenesis arise from the interactions occurring in populations of subunits, there are basically four types of perturbations available to us: (i) the modification of ambient environmental conditions, such as temperature, pH, salt concentrations, nutrient materials, etc.; (ii) *amputation*, the excision or removal from the system of some of the constituent subunits; (iii) *randomization*, the re-arrangement of some or all of the constituent subunits; and (iv) *hybridization*, the introduction into the system of new subunits from some external source. Each of these kinds of perturbations modifies the interactions between the subunits in some way, and hence gives rise to a new morphogenetic process which may be compared to the original one.

(c) From what we have already said, the main object of interest in all cases is the stability of form, and of the dynamical process which gives rise to form. Since the examples we have considered arise from interactions of substructures, we may hope to obtain an integrated view of stability questions applicable to all such processes, however much they may differ in the character of the subunits involved, or in the specific nature of the interactions between those subunits. It is indeed on this kind of premise that all of the subsequent lectures are based.

Before turning to a further development of this premise, however, it is advisable to consider another method by which we may seek an explanation for biological form and its generation. Indeed, no discussion of form and morphogenesis would be complete without such considerations. As we shall see in Lectures IV and V, they are in fact closely related to the main body of these lectures, although initially they seem quite different in character. Let us therefore see how they arise.

As noted at the outset, properties of form and pattern are the most readily perceived and universal aspects of organisms. They are at the root of the basic biological study of *taxonomy,* the classification of the biosphere in terms of morphological and morphogenetic attributes. Taxonomy was already a well-developed study in the time of the ancient Greeks, and was rigorized by Linnaeus in his classical works, dating from circa 1750. It was found that the totality of organisms, both modern and ancient, could be naturally grouped on the basis of their anatomical and developmental properties. The basic taxonomic unit is the *species,* defined partly in terms of the capacity of different members of the species to mate with each other and leave offspring.

Like any "natural" taxonomy, such as the Periodic Table of chemistry, the Linnean classification of organisms had many suggestive properties. There was found to be a gradation from the simplest organisms to the most complex, and there tended to be a continuity in these properties (especially when the fossil record was considered) which could hardly be fortuitous. One way of explaining these continuities was in terms of the idea of *evolution;* somehow, the more complex organisms, considered in terms of the classification scheme, were generated from the simpler ones, or *evolved* from them, over the course of time.

Since the basic taxonomic unit is the species, and since a species is characterized primarily in morphological and morphogenetic terms, the evolution of new taxonomic units represents, itself, a morphogenetic process, the creation of a new form from a pre-existing one.

Ideas of evolution also go back to the Greeks. But until a specific mechanism was proposed to explain how evolution could occur, the idea was more a philosophical speculation than a scientific hypothesis. It was Darwin's great contribution (first explicitly presented in his book, significantly titled, "The Origin of Species", in 1859) to propose such a mechanism.

The mechanism proposed by Darwin was called by him "natural selection"; it was empirically based on observations made by Darwin on artificially isolated populations, such as those trapped on islands, and on centuries of experience with animal and plant breeding aimed at enhancing desirable characteristics and suppressing undersirable ones.

Natural selection is concerned with the total functioning of a particular organism in a particular environment, as compared with the functioning of the other organisms with which it must compete and contend. This total functioning is conventionally described in terms of a single numerical parameter, called *fitness,* and is measured by a capacity of an organism to leave offspring to populate the subsequent generation (i.e. by the organism's *fecundity*). Intuitively, if the fitness of different organisms in a particular species varies, due to individual differences, and if the more fit organisms can transmit their characteristics to their offspring through a hereditary mechanism, then the entire population will become successively enriched in fit organisms. And since the biological functions on which fitness depends are intimately related to morphological and developmental characteristics, there will be a corresponding modification of form as well.

These evolutionary ideas on the generation of form have led to a number of theoretical approaches to the problem of biological form, which we shall now briefly describe. On the one hand, we find the idea that the successive variations of form arising from natural selection can be represented in purely geometric terms, in which we regard new forms as *transformations* of the initial one. This approach was most spectacularly developed by D'Arcy Thompson in his classical book (1917). He proceeded to embed appropriately chosen initial forms (taken two-dimensional for convenience) into a rectangular co-ordinate system, and showed how related forms could be obtained by relatively simple curvilinear transformations of those co-ordinates. More limited, but at the same time more explicit, ideas were being developed in connection with the variously called *allometric law*, or *law of heterogony*, or *law of heterauxesis* (see Huxley, 1932). If a pair of magnitudes x, y connected with a set of evolutionarily (or otherwise) related forms are measured, it is very often found that these magnitudes are themselves related by a power law

$$y = \alpha x^{\beta} .$$

Clearly, sets of such allometric relationships in a set of forms generate transformations of the type considered by D'Arcy Thompson, and conversely. Such work was carried still further by such authors as Lambert and Teissier (1927), who argued that the transformations relating different organisms must be considered as *space-time transformations*, i.e. they must relate whole developmental pathways, and not merely static forms manifested at particular instants. These concepts were further extended to the chemical level, in a tentative way, by Needham (1934). All of these ideas have the interesting feature that they require related organisms to be *models* of each other, in a precise and powerful sense; knowing the properties of any one organism, and knowing the appropriate transformation relating this organism and a second, we can find all the properties of the second organism. We shall meet with such ideas again in Lecture IV; it is interesting to note that they appear to have been largely forgotten by biologists, not because of any lack of intrinsic merit, but because of the concomitant growth of the exciting new techniques which led to molecular biology, and because of the lack of correspondingly powerful mathematical techniques by which their consequences could then be explored. An interesting modern version of these ideas of *biological similarity* can be found in Gunther (1975).

A related but rather different line of thought bearing on biological form, suggested by evolutionary considerations, can be found in the circle of ideas collectively called *optimal design*. Briefly, if natural selection operates to continually improve biological functions, then in situations in which selection operates strongly, we can expect to find these functions carried out in a fashion which is optimal, according to traditional design criteria known from engineering and physics.

Such ideas relate to considerations of biomechanics, on the one hand, aspects of which were already considered by Leonardo da Vinci and Galileo (cf. Haldane, 1956), and which lead at the same time to another way of viewing biological similarities (Stahl, 1961). On the other hand, optimality considerations can relate directly to the theory of transformations, since the trajectories of forms arising from successive transformations applied to an initial form can be looked on as the minimal curves of an appropriate variational principle (cf. Rosen,1967), which embodies the character of the selection process leading to a particular evolutionary trajectory. The entire approach to form through considerations of optimal design is due to Rashevsky (1960).

It should be noted that the ideas of similarity and biological transformation discussed above may be applied to any sequence of related forms. In particular, they can be applied to the successive forms of a single developmental process, and indeed have been widely studied in this context (cf. Richards and Kavanagh, 1945; Medawar, 1945). The formal homology between developmental and evolutionary morphogenetic processes bears a suggestive relationship to the idea of *recapitulation,* that *ontogeny* (the development of an organism from zygote to adult) recapitulates *phylogeny* (the sequence of evolutionary transformations leading to the species in question from the origin of life). For a stimulating review of these ideas, see deBeer (1940).

All of the approaches to form which we have outlined above are valid and important, and are directly related to the evolutionary underpinnings of modern biology. However, we feel intuitively that problems of form and morphogenesis must be attacked on other grounds as well. Basically, the approaches mentioned so far take form as an undefined primitive term, to be manipulated by *a priori* transformations externally applied. We feel that such transformations ought rather to be consequences and corollaries of a theory of morphogenesis based on underlying biological and physical mechanisms responsible for developmental processes. Such a theory must proceed by analyzing biological systems into appropriate substructures, and view the generation of pattern and form as proceeding from the interactions of these substructures. This is indeed the viewpoint most naturally arising out of the experimental and empirical considerations with which this lecture began. The remainder of these lectures will consider the problem of how this may best be accomplished.

REFERENCES

Anfinsen, C.B. (1968) Spontaneous Formation of Three Dimensional Structure of Proteins. In *The Emergence of Order in Developing Systems*, M. Locke, ed. (New York: Academic Press, Inc.).

deBeer, G.R. (1940) *Embryos and Ancestors*. Oxford: Oxford University Press, Inc.

Bonner, J.T. (1967) *The Cellular Slime Molds*. Princeton: Princeton University Press, Inc.

Briggs, R. and King, T.J. (1959) Nucleocytoplasmic Interactions in Eggs and Embryos. In *The Cell, Volume I*, J. Brachet and A.E. Mirsky, eds. (New York: Academic Press, Inc.)

Caspar, D.L.D. and Klug, A. (1962) Physical Principles in the Construction of Regular Viruses. *Cold Spring Harbor Symposia on Quantitative Biology, Volume XXVII*.

Child, C.M. (1941) *Patterns and Problems of Development*. Chicago: University of Chicago Press.

Crick, F.H.C. and Watson, J.D. (1956) The Structure of Small Viruses. *Nature, 177*, 473-475.

Curtis, A.S.G. (1967) *The Cell Surface*. New York: Academic Press, Inc.

Dalcq, A.M. (1938) *Form and Causality in Early Development*. Cambridge: Cambridge University Press.

Faber, J. (1971) Vertebrate Limb Ontogeny and Limb Regeneration: Morphogenetic Parallels. In *Advances in Morphogenesis, 9*, 127-147.

Gunther, B. (1975) *On Theories of Biological Similitude*. Fortschritte der Experimentellen und Theoretischen Biophysik, Bank 19, Leipzig: Georg Thieme Verlag.

Haldane, J.B.S. (1956) On Being the Right Size. In *The World of Mathematics, Volume 2*, J.R. Newmann, ed. (New York: Simon and Schuster, Inc.)

Hämmerling, J. (1946) Neue Untersuchungen über die physiologischen und genetischen Grundlagen der Formbildung. *Naturwissenschaften, 33*, 337-342, 361-365.

Huxley, H.E. (1967) The Fine Structure of Striated Muscle and its Functional Significance. In *Harvey Lecture Series, Volume 61*, (New York: Academic Press, Inc.)

Huxley, J. (1932) *Problems of Relative Growth*. London: Methuen and Company, Ltd.

Klug, A. (1969) Point Groups and the Design of Aggregates. In *Symmetry and Function of Biological Systems at the Macromolecular Level,* A. Engström and B. Strandberg, eds. (New York: John Wiley and Sons, Inc.)

Lambert, R. and Teissier, G. (1927) Theorie de la Similitude Biologique. *Annales de Physiologie et de Physicochimie Biologique, 2,* 212-246.

Levine, M. (1971) The Assembly of Phage from Subunits. In *Topics in the Study of Life,* 32-38. (New York: Harper and Row Publications, Inc.)

Locke, M. (1967) The Development of Patterns in the Integument of Insects. In *Advances in Morphogenesis, 6,* 33-88.

Medawar, P.B. (1945) Size, Shape and Age. In *Essays on Growth and Form,* W.E. leGros Clark and P.B. Medawar, eds. (Oxford: Oxford University Press, Inc.)

Mintz, B. (1967). Gene Control of Mammalian Pigmentary Differentiation, I. Clonal Origin of Melanocytes. *Proceedings of the National Academy of Sciences of the United States of America, 58,* 344-351.

Needham, J. (1934) Chemical Heterogony and the Ground-Plan of Animal Growth. *Biological Reviews, 9,* 79-109.

Rashevsky, N. (1960) *Mathematical Biophysics, Volume II.* New York: Dover Publications, Inc.

Richards, O.W. and Kavanagh, A.J. (1945) The Analysis of Growing Form. In *Essays on Growth and Form,* W.E. le Gross Clark and P.B. Medawar, eds. (Oxford: Oxford University Press, Inc.)

Rosen, R. (1967) *Optimality Principles in Biology.* London: Butterworths Publishing, Inc.

Scheraga, H.A. (1969) Calculations of Conformations of Polypeptides from Amino Acid Sequences. In *Symmetry and Function of Biological Systems at the Macromolecular Level,* A. Engström and B. Strandberg, eds. (New York: John Wiley and Sons, Inc.)

Singer, M. (1968) Some Quantitative Aspects Concerning the Trophic Role of the Nerve Cell. In *System Theory and Biology,* M.D. Mesarovic, ed. (New York: Springer-Verlag New York, Inc.)

Stern, C. (1956) *Cold Spring Harbor Symposia on Quantitative Biology, Volume XXI,* 375-382.

Spooner, B.S. (1975) Microfilaments, Microtubules, and Extracellular Materials in Morphogenesis. *Biosciences, 25,* 440-451.

Stahl, W.R. (1961) Dimensional Analysis in Mathematical Biology. *Bulletin of Mathematical Biology, 23,* 355-376.

Tartar, V. (1961) *The Biology of Stentor.* New York: Pergamon Press, Inc.

Trinkaus, J.P. (1969) *Cells Into Organs.* New Jersey: Prentice-Hall, Inc.

Thompson, D.W. (1917) *On Growth and Form.* Cambridge: Cambridge University Press.

Waddington, C.H. (1954) *Principles of Embryology.* London: Allen and Unwin, Ltd.

Waddington, C.H. (1962) *New Patterns in Genetics and Development.* New York: Columbia University Press.

Warren, K.B. (ed.) (1967) *Formation and Fate of Cell Organelles.* New York: Academic Press, Inc.

Weiss, P. (1939) *Principles of Development.* New York: Holt, Rinehart and Winston, Inc.

Weiss, P. (1968) *Dynamics of Development: Experiments and Inferences.* New York: Academic Press, Inc.

Wigglesworth, V.B. (1945) Growth and Form in an Insect. In *Essays on Growth and Form,* W.E. le Gross Clark and P.B. Medawar, eds. (Oxford: Oxford University Press, Inc.)

Wolpert, L. (1969) Positional Information and Pattern Formation. In *Towards a Theoretical Biology: Volume 3, Drafts,* C.H. Waddington, ed. (Chicago: Aldine-Atherton, Inc.)

Lecture II

As we have seen, biological structures can be analyzed into assemblies of subunits of various types. On the one hand, such an assertion can be regarded as trivial; the subunits may simply be taken to be the atoms and molecules of which all matter is composed. The remarkable fact is that this idea of the assembly of organic structures out of subunits persists at higher levels of organization. For instance, the biologically important macromolecules are all polymers, built out of a small family of monomeric subunits. Thus, proteins are essentially polypeptides, built out of approximately 20 amino acids; the nucleic acids are polymers built out of four or five kinds of nucleotides; the polysaccharides are built out of a similar small number of elementary sugars. As noted earlier, many biological organelles are macromolecular complexes, each unit of which is itself a polymeric macromolecule. The cell theory of organisms has already been mentioned as a further illustration of this idea, at a still higher level.

The organization of biological systems out of subunits, which are themselves assemblies of still smaller subunits, down to the atomic level, is one of the features which gives biological structures their pronounced hierarchical character. Furthermore, the universality of this mode of organization in biology suggests that there is something particularly efficient about such hierarchical structures. In the present lecture, we shall review some formal aspects of this circle of ideas, which comprise the bulk of the theoretical literature on morphogenesis.

Let us first consider how one goes about generating complex structures from a family of elementary units. There are basically only two ways in which this can be done, depending on the properties of the units. It may be, for instance, that the units possess no particular affinities for each other, and no particular intrinsic dynamics, like a pile of ordinary bricks. In order to create a structure out of such units, an external agency must be invoked. This agency, which we may generally call a *template,* must in some sense contain the information regarding the final structure to be created. In general, each part of the template must have a specific affinity for the appropriate unit. Thus, the differential affinities of different regions of the template induce a spatial order into the subunits, and hence a specific form is created, which mirrors the properties of the template. This mode of construction, it will be recognized, is the qualitative picture which has been developed to explain the synthesis of biologically important macromolecules, such as the proteins and the nucleic acids. Moreover, besides the information contained in the template, there is information contained in all the accessory structures, such as catalysts and activators, required in order to allow a template to organize a family of inert monomeric units into a specific stable structure. Since the information required for creating such a structure from subunits must be provided from sources outside of the subunits themselves, we could not hope to predict any particular final form from a study of the subunits in isolation.

On the other hand, the units themselves may possess intrinsic affinities for each other, and hence a definite tendency to organize *themselves* into more complex structures, independent of any external source of information. In this case, in effect, the "information" required to generate more complex structures from the units inheres in the properties of the units. From a knowledge of these properties, then, we should be able to predict the characteristics (i.e. the form) of the kinds of structures generated from them. It will be recognized that such populations of units, possessing inherent differential affinities for each other, provide the prototype for that class of processes which have been called *self-organizing,* or *self-assembling*.

Within the class of self-organizing systems, we can recognize two further cases. On the one hand, the properties of the individual units can remain constant in time. This is the situation, for instance, in the spontaneous reassembly of a tobacco mosaic virus, or in the cell-sorting situations described in the previous lecture. But we can also imagine that the intrinsic properties of the subunits can themselves change with time, either autonomously, or in response to environmental influences, or both. This situation obtains, for instance, in developmental processes, in which the properties of the constituent cells become progressively modified in time through a process of *differentiation*. Such changes in the properties of the subunits may reflect themselves, of course, in modifications of the affinities of the units for each other, and a corresponding modification in the structures arising from assemblages of these subunits.

So far, we have considered a fixed set of subunits, which could generate structures through differential affinities for each other, and through differentiation within the subunits. But especially in biology, we must consider also the case in which the population of subunits is itself changing in time. This can occur either through external sources and sinks for the subunits, or through the replication and death of subunits. This kind of process, superimposed on differentiation, will allow new populations of units to appear and to grow, and old populations to decline and disappear.

The mechanisms we have just described comprise what we shall call the three *fundamental morphogenetic mechanisms* which can play a role in the generation of pattern and form in populations of active units. We may characterize them succinctly as follows:

1. *Morphogenetic movements*. By this term, we refer to processes which generate form and pattern by spatial rearrangements in a population of units, without change either in the properties of the units, or in their numbers.

2. *Differentiation*. By this, we mean forms or patterns arising from differential changes occurring in the units themselves, without any change in the number of units or in their spatial organization.

3. *Differential birth and death*. By this, we mean differential increases and decreases in the numbers of units of our family, without any change in their properties.

The first explicit formulation of this kind of analysis of form-generating processes seems to be due to Bonner (1952). Clearly, any real morphogenetic process is a combination of these three different elementary mechanisms. And equally clearly, these three elementary mechanisms cannot be regarded as entirely independent of each other. Nevertheless, it has proved to be an effective strategy to characterize each of these elementary morphogenetic mechanisms in abstract, mathematical terms, and to assess the *morphogenetic capabilities* of each of these kinds of mechanisms considered in isolation. That is, we shall attempt to discover what kinds of forms and patterns can, and cannot, be conveniently generated by morphogenetic movements alone, by differentiation alone, and by differential birth and death alone. In this way, we can hope to approach any real morphogenetic process armed with insights regarding the capabilities of the elementary morphogenetic mechanisms to account for that process. Indeed, the isolation of the elementary mechanisms amounts to a conceptual counterpart, in dynamical terms, of the decomposition of morphogenetic structures themselves into constituent subunits. That is, we have generated a mode of analysis, whose ramifications are still in the process of being actively explored.

Analysis of morphogenesis in these terms has several kinds of goals, which should be stated explicitly. These include:

1. A capacity to predict, from suitable initial information regarding the units, what the final form generated by those units will be like. Part of the analysis involves a determination of what constitutes "suitable initial information".

2. A capacity to predict the dynamical process (i.e. the trajectory) whereby the final form is generated from the initial array of units.

3. A capacity to predict how the system will respond to perturbations, especially the experimentally important perturbations of amputation, randomization and hybridization. This involves the assessment of the *stability* of form, and of the morphogenetic process itself, to such perturbations. It also explicitly embodies the hypothesis that *the process which gives rise to a form must also stabilize that form.* Indeed, as we shall see, questions of generation of pattern and form are inextricably bound up with questions of maintenance and stability.

A. The Mathematical Representation of Morphogenetic Movements

In this section we consider the morphogenetic capabilities of a population of units (which may be cells, or molecules, or molecular aggregates) which (a) possess differential affinities for each other, and (b) possess motility, which permits them to explore alternative local environments. We shall assume that the properties of the subunits, and also their numbers, do not change during the course of the analysis. It will be recognized that these assumptions are satisfied in a number of important cases of biological self-assembly which were discussed in Lecture I, at a variety of morphogenetic levels; the assembly of micro-filaments and microtubules; the assembly of viruses; the sorting-out of embryonic cells; the folding of polypeptides; as well as to purely inorganic processes collectively called phase separations.

We proceed by constructing a formal framework in which the basic conditions of differential affinities and motilities of the subunits can be given a concrete representation. Thus, let us imagine a tessellation of a region of the plane, as shown in Figure II.1.

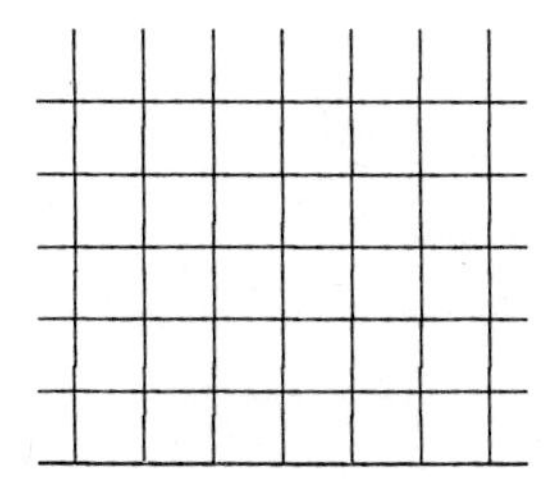

Figure II.1

Each square in the tessellation will contain a unit from our population. If for simplicity we consider the situation where the population consists of two types of units (designated A and B respectively, with relative numbers N_A, N_B respectively) then an assignment of a unit to a square of the tessellation produces what we shall call a *pattern*. Several typical patterns, with $N_A = 10$, $N_B = 15$, are shown in Figure II.2.

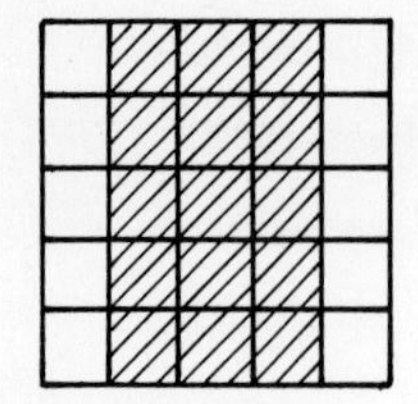 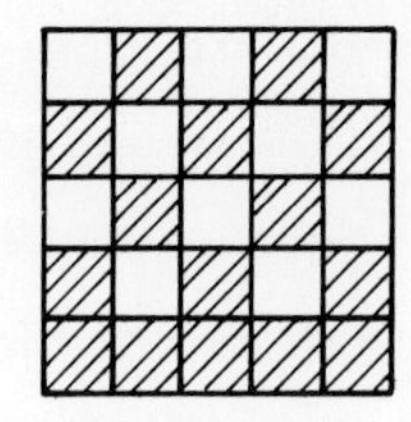 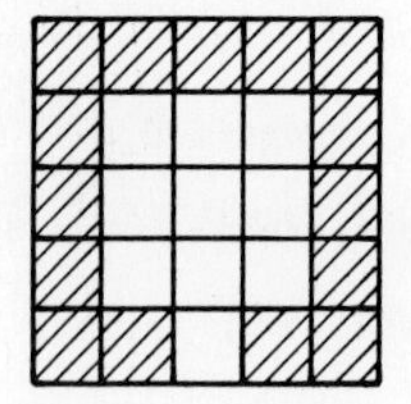

Figure II.2

We shall suppose that adjacent units in the tessellation can interact with each other across their common side. In any pattern of the type we are considering, three kinds of interactions can occur: (a) two B-units lying in adjacent squares can interact; (b) two A-units lying in adjacent squares can interact; (c) an A-unit and a B-unit lying in adjacent squares can interact. To each such interaction, we can assign an "affinity", i.e. a non-negative number which measures the interaction between the units across the common edge. Thus, in our case, there will be three such affinities, denoted respectively by λ_{BB}, λ_{AA}, λ_{AB}. In general, with r kinds of units, there will be $\binom{r+1}{2}$ affinities. Moreover, given any particular pattern, the number of edges across which interactions of each type are manifested is determined by the pattern. We can thus define a "total energy" associated with the pattern by

$$E(P) = N_{AA}\lambda_{AA} + N_{BB}\lambda_{BB} + N_{AB}\lambda_{AB}$$

where N_{AA}, N_{BB}, N_{AB} are the numbers of edges in the pattern separating a pair of A-units, a pair of B-units, and a pair consisting of one A-unit and one B-unit. In this way, we associate a number $E(P)$ with a pattern P; $E(P)$ is reciprocally related to the "free energy" of the pattern, and represents a mathematical embodiment of the hypothesis of the differential affinities of the units.

Let us now see how the requirement of motility can be incorporated into this formalism. Intuitively, we would expect that, in a suitable small time interval, such motility would be manifested by interchanging the units occupying neighboring squares in the tessellation. In general, given a pattern P, we shall say that a pattern P' which can be obtained from P by the interchange of one or more adjacent units is a *neighboring pattern* of P. Thus, each pattern P determines a set $N(P)$ of neighboring patterns, reachable from P by one or more interchanges of adjacent units. With each of the patterns $P' \in N(P)$ we can associate an "energy" $E(P')$ in the manner described above. We choose, in $N(P)$, that pattern P' for which the associated $E(P')$ is *minimal* over $N(P)$, and we shall say that the pattern P will *generate* the pattern P' in an appropriate time interval.

We can repeat this procedure with the new pattern P'; this pattern determines a set $N(P')$ of neighboring patterns; in this set we choose a pattern P'' for which $E(P'')$ is minimal over $N(P')$. Proceeding in this fashion, we obtain a sequence of patterns

$$P \rightarrow P' \rightarrow P'' \rightarrow \ldots$$

This process terminates when we reach a pattern $P^{(n)}$ such that, for any pattern $P \in N(P^{(n)})$, $E(P) \geq E(P^{(n)})$. This terminal pattern $P^{(n)}$ can be regarded as the pattern of "minimal free energy" generated by the initial pattern.

The mathematical formalism we have described seems paradigmatic, at least at the phenomenological level, for the investigation of all cases of self-assembly arising from differential affinities and motility. It has been investigated in detail, mainly through simulation techniques, for a wide variety of situations (cf. the recent volume edited by Mostow (1975), which reproduces many of the original articles). In particular, the following generalizations have been considered: (a) variation of the tessellation geometry; (b) introduction of anisotropy into properties of the constituent units; (c) variation of the "motility" rules, such as allowing exchange of units possessing an adjacent corner as well as units possessing an adjacent side, or allowing transition sequences which do not strictly decrease monotonically in free energy.

In addition to verifying the experimental facts, particularly those associated with cell-sorting, a variety of other interesting features have emerged from these investigations:

(a) Results more closely constant with experiment are obtained when the units are allowed to obtain "information" regarding the contents of squares which are not adjacent to them in the tessellation. This is represented by allowing a unit to make a locally unfavorable move, provided that by so doing the unit can reach an energetically more favorable situation than it otherwise could. This introduces an element of "dynamic programming" into the sorting algorithms, and is, moreover, consistent with experimental indications that, at the cellular level, the units do indeed obtain, and act on, information about regions with which they are not in direct contact.

(b) As represented in the formalism, pattern generation is a holistic process. That is, energetically unfavorable structures can be preserved in such patterns, provided that the "energetic cost" for them is borne by other regions of the pattern. If the energetically unfavorable region is separated from the remainder of the pattern, then that region will re-organize and the structure will be lost.

(c) Since pattern-generation through cell movements will, as we have seen, occur in any family of cells, the forces responsible for cell sorting must be consistent with other kinds of morphogenetic mechanisms in embryological development. If this were not so, the sorting mechanism would tear apart structures arising through differentiation and birth-and-death processes. This is undoubtedly why randomized structures, such as lung and kidney, taken from early embryos, will reassemble through cell sorting, even though they were not originally generated in this fashion. Indeed, as these organs develop further, mainly through further differentiation mechanisms, the capacity to reconstitute them through sorting-out is progressively lost; commensurate with this further development, other modes of stabilizing the pattern (e.g. through extracellular matrices of collagen and other materials, or intracellular "tight junctions") are found to appear.

Thus far, most of these studies on pattern generation have emphasized stability properties. Much work remains to be done on the actual kinetics in generating particular patterns, as an explicit function of the affinities. This kind of study would throw light on developmental abnormalities arising from changes in cell affinities (for instance, that studied by Ede (1973), affecting limb morphogenesis in the chick), as well as on important medical problems such as invasiveness and metastasis by tumor cells.

In connection with the application of this paradigm to problems of protein folding, we must mention the pioneering work of Pyatetskii-Shapiro and his co-workers (cf. Mostow, 1975). In this work, the problem involved finding motility rules which would move a population of motile, differentially adhesive units, constrained now to lie on a polygonal curve (i.e. forming a polymeric "molecule"), to a circular configuration. The bearing of this kind of approach on the problem of molecular conformations is obvious, and provides an alternative to more physical or structural approaches, as embodied especially in the work of Scheraga (1969), and outlined in Lecture I.

B. The Mathematical Representation of Differentiation

We will now turn to the problem of mathematically representing those morphogenetic processes which involve modifications in the intrinsic properties of a family of units (cells), without change in their number or position.

The prototypic problem of differentiation arises from the creation of polarities, or assymmetries, or gradients, in previously homogeneous or symmetric situations.

The apparently autonomous generation of gradients and assymetries in initially homogeneous systems has always been one of the most puzzling, as well as the most ubiquitous, features of biological activity. The idea of a system spontaneously proceeding from homogeneity to inhomogeneity, or from "disorder" to "order", seems to contradict elementary thermodynamic considerations regarding entropy. Therefore, for many years, the attention of those concerned with understanding the basis for differentiation was drawn mainly to finding plausible sets of assumptions under which this kind of behavior would be manifested.

The first attempts in these directions concerned the behavior of open systems, i.e. systems which could exchange matter and energy with their environment. It was noticed that only for closed, isolated systems did the physical arguments regarding maximization of entropy hold. Moreover, the asymptotically stable steady states of an open system automatically exhibited regulatory properties often found in developing systems. Further, none of these steady states need correspond to a homogeneous situation. All of these aspects were discussed in the early literature, especially by von Bertalanffy (1953). Simultaneously, within physics itself, attention came to be directed at the thermodynamic properties of open systems, in the work of men like Onsager (1931), de Groot (1962), and Prigogine (1955), without, however, any immediate influence from biology.

Apparently the earliest explicit mechanism, based on physically plausible assumptions, for the generation of polarities and gradients was developed by Rashevsky (1940), on the basis of his investigations into the mechanical forces produced by diffusion gradients. He considered a spherical cell, in which sources for some chemical reaction are distributed uniformly. On the basis of his diffusion-drag-force theory, particles suspended in the cell will tend to accumulate in regions of low concentration of the reaction products. If we now further assume that such particles are able to *inhibit* the reaction, then the net effect of the forces in this system will be to *enhance* the gradients. In effect, the dynamics of this system exhibit a positive feedback loop, which cumulatively magnifies any initial gradient. Rashevsky noticed that if the entire system was initially spherically symmetric, then it would remain so forever; however, the smallest departure from spherical symmetry would likewise be magnified by the positive feedback loop. Ultimately, all of the inhibitory particles would come to lie in that hemisphere which initially had more of them, leaving the other hemisphere with all of the reactants and products. Thus a polarity would automatically be established in such a system, and this polarity would be stable to randomization; the gradient would be re-established no matter how the contents of the cell were stirred, or how the cell was divided.

Thus, Rashevsky's idea was the following: to construct a dynamical open system, whose homogeneous or symmetric steady states were unstable, so that any perturbation away from those states would be carried by the system dynamics to a new steady state, necessarily inhomogeneous. A positive feedback loop, created by the dynamics, would serve to magnify any initial departure from homogeneity. All subsequent treatments of pattern generation through differentiation are variations of this basic mechanism.

The same idea was presented in simpler form by Turing (1952). Since Turing's paper has been the point of departure for many of the recent theoretical developments, we will spend some time reviewing his ideas.

Let us suppose that the state of an abstract cell can be specified at an instant of time t_0 by a single number $s(t_0)$, which will represent the concentration of a chemical substance x. Such a substance will be called a *morphogen*. Suppose that morphogen is supplied to the cell from an external source at a constant rate S, and is being destroyed at a rate proportional to its concentration. Then the dynamics of the system is expressed by the single differential equation

$$\frac{dx}{dt} = -ax + S.$$

This system has a single steady state $x^* = S/a$, and this steady state is asymptotically stable.

If we consider two such cells, whose states are specified by the morphogen concentrations in each of them, then the two-cell system is governed by the equations

$$\frac{dx_1}{dt} = -ax_1 + S$$

$$\frac{dx_2}{dt} = -ax_2 + S.$$

It is readily verified that this degenerate two-dimensional situation possesses only one steady state, given by $x_1{}^* = x_2{}^* = S/a$, and that this steady state is a stable node.

Finally, we allow the two cells considered above to interact through diffusion of morphogen. We suppose that this interaction is proportional to the difference in morphogen concentration between the cells, according to the system of equations

$$\begin{aligned} \frac{dx_1}{dt} &= -ax_1 + S + D(x_1 - x_2) \\ \frac{dx_2}{dt} &= -ax_2 + S + D(x_2 - x_1) \end{aligned} \qquad \text{(II.1)}$$

These equations are still linear, and still possess a unique steady state at $x_1 = x_2 = S/a$. We may solve them by introducting the new co-ordinates

$$\begin{aligned} u_1 &= x_1 + x_2 - 2S/a \\ u_2 &= x_1 - x_2 \end{aligned}$$

The equations of the system then become

$$\frac{du_1}{dt} = -au_1$$

$$\frac{du_2}{dt} = (2D - a)u_2$$

We can see immediately that one of the eigenvalues of the system has become a function of the diffusion constant D, which measures the coupling between the cells. If $D < a/2$, the homogeneous steady state remains a stable node. However, when $D > a/2$, that same steady state becomes a *saddle point,* with trajectories shown in Figure II.3.

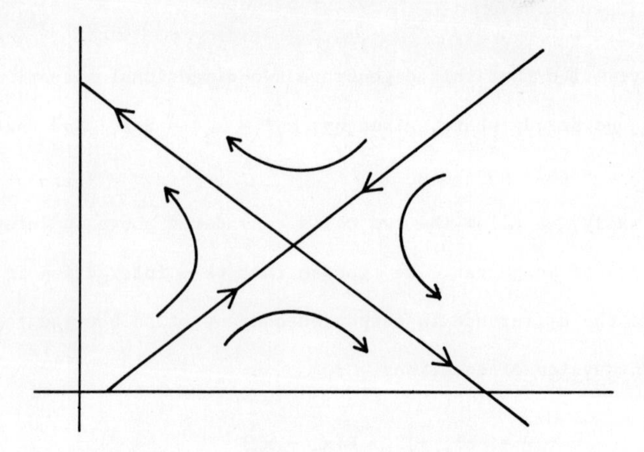

Figure II.3

Since the original state variables x_1, x_2 represent concentrations, they cannot become negative; hence the dynamics must stop when a trajectory intersects one of the axes.

The similarity of this scheme to the idea of Rashevsky is now apparent. The homogeneous steady state is unstable; the least perturbation away from it will be magnified by the system dynamics, in such a way that the system will stop when all the morphogen has been removed from one of the cells, and "pumped" into the other. Here again, a gradient or asymmetry will develop as an automatic consequence of the system dynamics.

The simple one-morphogen situation treated above has the advantage of being easy to visualize geometrically. However, to generate the required behavior, we required an *active* coupling between the cells, rather than passive diffusion. If we allow a two-morphogen situation, as Turing did in his original paper, then the same behavior arises with only passive diffusion. In this case, it will be noticed that the same mechanism which creates or generates a pattern also serves to create an *active transport;* this feature seems to have been generally overlooked.

The mechanism proposed by Turing can be generalized in a large number of different ways. For one thing, we can consider patterns formed in any family of cells interacting by diffusion. The geometry of the cellular pattern is mirrored in the dynamical equations through the diffusional coupling, which only occurs between neighboring cells. Thus, in the case of a ring of cells, as shown in Figure II.4, and a single morphogen, we can

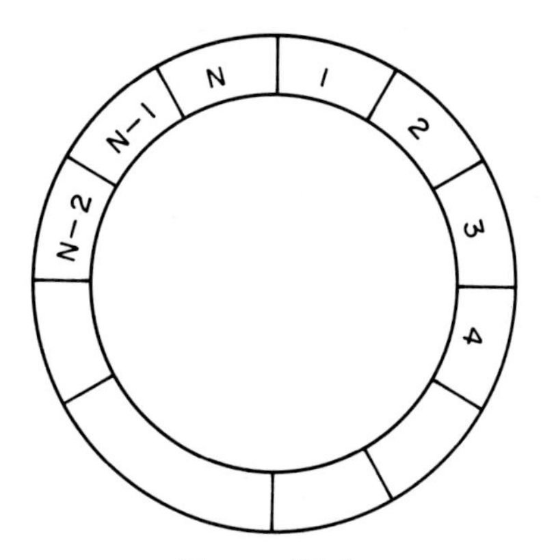

Figure II.4

immediately write down the equation

$$\frac{dx_i}{dt} = -ax_i + S + D(x_{i+1} + x_{i-1} - 2x_i)$$

with the identification $x_{N+1} = x_1$.

We can also generalize the analysis to any number of morphogens. A full treatment of the general situation, in which any arbitrary cellular geometry and an arbitrary number of morphogens are involved, has been analyzed in detail by Othmer and Scriven (1971).

The fact that the Rashevsky-Turing ideas involve chemical reactions and diffusion only, and that they can be responsible for behavior which apparently contradicts the behavior manifested by closed isolated systems, has recently led to some fundamental approaches to these systems by Prigogine and his co-workers (cf. Prigogine and Glansdorff (1971)) from the viewpoint of irreversible thermodynamics. The automatic passage of such systems from a symmetric to an asymmetric state has led to their being termed *symmetry-breaking instabilities*. We do not have the space to review this extensive and important work, except to note it as an instance in which biological phenomena have caused physicists to substantially enlarge and revise the scope of their previous work.

It should also be noticed that the problem of determining the distribution of morphogens in a geometric array of cells has many close relationships with other "network" theories of biological activity, particularly with those arising from networks of excitable elements. The problem there is indeed one of differentiation: which elements shall be excited, and which shall be quiescent, when a network of specific geometry is placed, through external excitation, in some initial state? Indeed, as we have shown (Rosen, 1968), we can convert a morphogen system into a network of excitable elements by simply modifying the coupling between the elements, e.g. by multiplying the diffusion interaction $D(x_1 - x_2)$ between cells by a step-function which embodies the idea of a threshold. This provides another example of how systems of quite different structure and behavior can be closely related on formal terms; we shall go more deeply into this kind of relationship in Lectures IV and V. These remarks should also be kept in mind when we discuss network theories of gene activation in Lecture III.

As a final instance of the class of ideas we have been developing here, we must mention the Keller-Segel (1970) model for the aggregation of slime molds. In this analysis, the basic variables are also densities and concentrations, although they are interpreted not in chemical terms, but in terms of the positions of cells in a population.

In greater detail, let us recall from Lecture I that, under conditions of starvation, free-living individual amoebae begin to aggregate to form slugs, which then differentiate further. The individual cells are chemotactic, i.e. move along concentration gradients of a substance (*acrasin*) produced by the cells themselves. The cells are also able to degrade this substance through a specific enzyme, acrasinase. If $a(x,y,t)$, $\rho(x,y,t)$ respectively represent the density of amoebae on a plane surface and the concentration of acrasin on this surface, then Keller and Segel proposed the following equations

$$\frac{\partial a}{\partial t} = -\nabla\cdot(D_1\nabla\rho) + \nabla\cdot(D_2\nabla a)$$

$$\frac{\partial \rho}{\partial t} = -K(\rho)\rho + af(\rho) + D\rho\nabla^2\rho$$

for the dynamics of this system. The first equation contains two terms for the amoeba density, one arising from the chemotaxis, and the other from random diffusion of the cells. The second equation contains a synthesis term for acrasin, a term for its destruction through acrasinase, and a term representing diffusion. Keller and Segel then assume the existence of a homogeneous steady state, i.e. one for which

$$a(x,y,t) = a_0 \qquad \rho(x,y,t) = \rho_0$$

and investigate its stability through standard linearization techniques. They find that this homogeneous steady state can become unstable; such an instability will be manifested by specific aggregations of the cells. For fuller details, we refer to the original paper (Keller and Segel, 1970).

One final remark may be made in connection with the Rashevsky-Turing ideas, namely that the Rashevsky-Turing mechanisms provide a means for understanding *differentiation*, which we have taken as one of the fundamental morphogenetic mechanisms, in terms of chemical reactions and diffusion. The reactive and diffusive processes themselves pertain to an underlying microscopic description of the units, in terms of the molecules of which those units are composed. But at that molecular level, *reaction* (i.e. the conversion of reactant molecules into product molecules) is the analog of birth and death at the cellular level; *diffusion* is the molecular analog of morphogenetic movement at the cellular level. This observation has several interesting corollaries: (a) the hierarchical organization of biological systems, which was mentioned earlier, has the property that the morphogenetic mechanisms arising at any level are consequences of those same basic mechanisms operating at lower levels; (b) there is an exact analogy between the formalisms appropriate to describe these morphogenetic mechanisms; this analogy makes a single abstract formalism applicable at all levels; (c) in particular, the combination of cell movement and birth and death, which has not yet been extensively attempted, can be guided by the much better-developed theory of reaction-diffusion systems. The ability to treat many different structural levels in terms of a common formalism provides a most important conceptual unification of biological processes, different from that commonly offered in reductionistic approaches to biology; we shall discuss the basis for such unifications more extensively in Lectures IV and V below.

Although the Rashevsky-Turing mechanisms can generate a large number of patterns on the basis of the ubiquitous phenomena of chemical reactions and diffusion, they have been criticized on several grounds. Most significant is the criticism of Wolpert (1969), who in effect has argued that, in many-celled systems, the nature of the final pattern generated by the system will depend crucially on the direction of the original (random) displacement from homogeneity. Thus, he argues, the Rashevsky-Turing systems cannot account for many of the important regulatory phenomena in real morphogenesis. However, it is possible to formally overcome such arguments, essentially by allowing the cells themselves to be anisotropic. By such means, we can, for example, convert the two-cell system described by equations (II.1), whose trajectories are shown in Figure II.3, to a system whose trajectories are shown in Figure II.5.

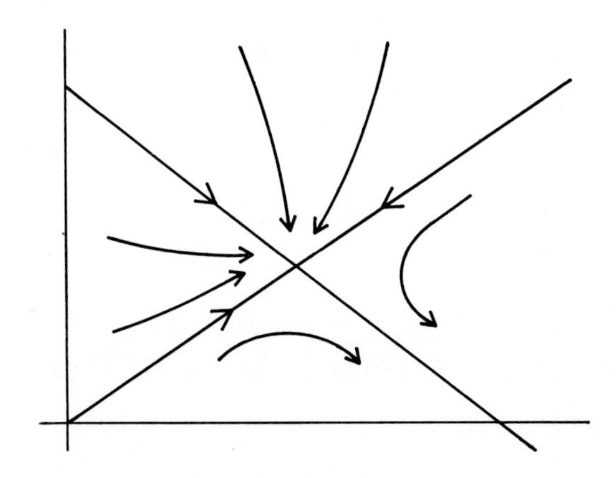

Figure II.5

In this system, a random displacement in a direction above the line $u_2 = 0$ will simply return to the homogeneous situation. Only a displacement below that line will provoke the instability. In this way, we can make sure that a Rashevsky-Turing system will always generate a unique pattern. The biological plausibility of such a mechanism, however, remains arguable. For further details, see Rosen (1971).

We might also mention the experiments of Gordon and Drum (1973), looking for such a mechanism in the generation of patterns in diatom shells. These authors argue that their results are not consistent with such an explanation of this specific morphogenetic situation.

We shall conclude this brief review of the mathematical representation of global differentiation mechanisms with the concept of "positional information", as developed by Wolpert (1968, 1969). According to Wolpert's analysis, the regulatory phenomena of embryology, as they are revealed especially by amputation and hybridization experiments, can be explained if one assumes that, in general, a cell in a developing system makes its "decisions" as to differentiation according to its relative position in the total aggregate. This kind of behavior is summed up by Wolpert in his paradigmatic "French Flag Problem". Here, we imagine a rectangular population of cells, each initially capable of three decisions (which we may call "red", "white" and "blue"). The pattern to be generated is the French Flag, with all cells in the leftmost third to differentiate into "blue" cells, all cells in the middle third to differentiate into "white" cells, and all cells in the rightmost third to differentiate into "red" cells. We wish the mechanism for this differentiation to have the property that, if a fragment of the pattern is removed, as shown in Figure II.6, so that the relative positions of the cells in the fragment is changed, the fragment will re-differentiate to form a "French Flag" in its own right.

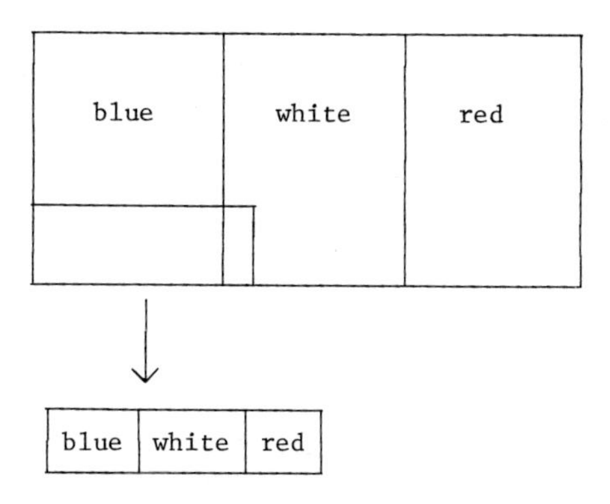

Figure II.6

One mechanism for generating such "positional information" has been presented by Goodwin and Cohen (1969). In this mechanism, a set of independent periodic cellular processes produces waves which propagate through the entire population. The phase differences between these waves provides a kind of co-ordinate system, according to which each cell in the aggregate is uniquely determined. The candidates for these periodic processes are to be found in ubiquitous genetic and epigenetic control mechanisms (see Lecture III). All of the schemes for generating "positional information", however, require some kind of externally established gradient.

Before we leave the subject of the dynamical representation of differentiation, mention must be made of the influential recent work of Rene Thom (1972). This work is of a rather different character than that discussed above; we will consider it, and its relation to the approaches we have already discussed, more extensively in Lecture V. Thom's work is primarily concerned with cataloguing the various morphogenetic possibilities arising from *bifurcations* in dynamical systems; a bifurcation represents a transition mode from one class of dynamical behaviors to another. For instance, we saw above, in our discussion of the Rashevsky-Turing mechanism, that the variation of the diffusion constant D appearing in equations (II.1), from a region for which $D > a/2$ to a region for which $D < a/2$, gives rise to such a qualitative dynamical change; the point $D = a/2$ is thus a bifurcation point for the system. The genesis of Thom's ideas is to be found in the earlier metaphorical picture, due to Waddington (1957), of differentiation in terms of a particle rolling down an "epigenetic landscape" - a surface of furrows and ridges; the upper parts of the "landscape" correspond to relatively undifferentiated situations, while the various furrows leading downward correspond to different specific fates for the cells moving along it. Bifurcations then correspond to regions in this landscape where several furrows emanate from a single region. A diagram of this situation is shown in Figure II.7.

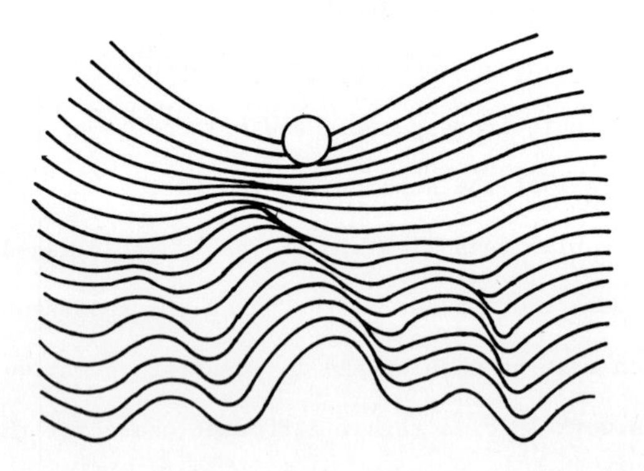

Figure II.7

The relation between this kind of "epigenetic landscape" and the state space of a dynamical system fibered into trajectories is quite clear. As noted, we shall provide a fuller discussion of Thom's work, which lies at a different level of generality from that considered above, in Lecture V.

C. Mathematical Representation of Differential Birth and Death

Thus far, we have considered situations in which the numbers of units in our population has not changed. Indeed, the numbers of cells of the various types we have considered has been an essential initial condition for our analyses. This is one of the serious drawbacks of the dynamical analyses we have considered; the change of a population of cells by even one cell requires an entirely separate analysis. Further, the relations between patterns generated in a family of N cells, and in a family of $N+k$ cells, e.g. in a Rashevsky-Turing system, must be established in an *ad hoc* manner, employing features of these patterns not intrinsic to the analysis itself.

The analysis of differential birth-and-death models, then, must necessarily proceed on a different basis than those we have considered heretofore. They do not fit naturally into an analytic framework, and are typically presented in terms of algorithmic processes, and hence, ultimately, in terms of the theory of automata. They proceed in terms of the external manipulation of *passive* units according to some rule or algorithm which, it is hoped, are actually implied by a more microscopic, intrinsic description of the units.

The earliest investigations into the patterns generated by growing populations of units were those of Eden (1956) and Ulam (1962). In these systems, a cell is represented by a structureless geometric object placed for convenience on a two- or three-dimensional grid. An algorithm is then specified for the placement of new forms on the grid, and for the removal of old forms, at successive time instants ("generations"), as a function of the pattern present at the previous instant. It is found that even very simple recursive rules of this type can generate complicated patterns; moreover, we typically cannot infer the algorithm which has generated such a sequence of patterns from a study of the sequence itself.

This kind of approach was carried a step further by Lindenmayer (1968a, 1968b, 1971). Lindenmayer introduced a kind of differentiation into his system as well, by allowing his units to exist in a number of distinct internal states. He stated his algorithms initially in automata-theoretic, and ultimately in linguistic, terms, by embodying them in a set of "production rules" whereby initial patterns could successively be transformed. For instance, let us suppose that each individual cell can be in one of two states, which we designate by a, b respectively. We also assume that the state of such a cell depends on the state of its immediate leftmost neighbor at the preceding instant, as well as on its own state at that instant. Suppose we start out with a string of cells, representing a filamentous organism, of the form

aabaab

The production rules are embodied in a "state transition" matrix of the form

		state of neighbor	
		a	b
state of cell	a	a	b
	b	bb	a

Here we see that the "reproduction" of a cell is embodied externally in these production rules. Applying them to the initial string displayed above, we obtain the following sequence:

t = 0: aabaab

t = 1: abbbabb

t = 2: baabbbab

etc.

With appropriate geometrical interpretations, Lindenmayer has shown that numerous patterns closely resembling real morphogenetic situations, especially in plants, can be constructed using such production rules. The grammars appropriate to these studies have been investigated by a number of authors (Arbib and Laing, 1972; Herman, 1971).

A somewhat similar approach has been developed by Cohen (1966). This work generates branching patterns from an initial "stalk" placed in a morphogenetic "gradient". Rules establish the probability of growth or branching at any point in the pattern, and specify the angle of branching. Small changes in the numerical values assigned to these various probabilities can lead to large changes in the pattern of arborization which results. (It is interesting to compare these patterns with studies of branching patterns made on biomechanical (Opatowski, 1945) or optimality (Howland, 1962; Rosen, 1967) grounds.)

These algorithmic or automata-theoretic approaches merge with those to be discussed in more detail in Lecture III. It will be noticed that the differential birth-and-death property is mirrored in the Lindenmayer models by external rules of production, which manipulate passive units, and which can replace single units by pairs of units. However, in recent years, a theory of autonomously "self-reproducing" units, couched in terms of the theory of automata, has appeared. It thus seems plausible to consider arrays of such inherently reproducing units, interconnected in various ways. In such a fashion, an algorithm, or set of production rules, governing differential birth-and-death arises automatically, instead of having to be posited in advance. Since such ideas are more directly connected with the activity of the genome than any we have considered heretofore, it is appropriate to deal with them in that context. A consideration of the genome, and its relation to the morphogenetic mechanisms we have discussed, is the substance of the remaining lectures; we therefore turn now to a review of the ideas of genetic control.

REFERENCES

Arbib, M.A. and Laing, R. (1972) Automata Theory & Development II. Preprint.

von Bertalanffy, L. (1953) *Biophysik der Fliessgleichgewichts.* Leipzig: Vierwig.

Bonner, J.T. (1952) *Morphogenesis.* Princeton: Princeton University Press.

Cohen, D. (1966) Computer Simulation of Biological Pattern Generation. *Quarterly Progress Report, Research Laboratory of Electronics, Massachusetts Institute of Technology,* 237-246.

Curtis, A.S.G. (1967) *The Cell Surface.* New York: Academic Press, Inc.

Ede, D.A. (1973) Cell Behavior in Embryonic Development. In *Physical Principles of Neuronal and Organismic Behavior,* M. Conrad and M.E. Magar, eds. (London: Gordon and Breach Science Publishers, Inc.)

Eden, M. (1956) A Probabilistic Model for Morphogenesis. In *Symposium on Information Theory in Biology,* N.P. Yockey, ed. (New York: Pergamon Press, Inc.)

Goodwin, B.C. and Cohen, M.H. (1969) A Phase-Shift Model for the Spatial and Temporal Organization of Developing Systems. *Journal of Theoretical Biology, 25,* 49-107.

Gordon, R. and Drum, R. (1973) Pattern Formation in Diatom Shells: A Test of Turing's Theory of Morphogenesis. Preprint.

de Groot, S.R. and Mazur, P. (1962) *Non-Equilibrium-Thermodynamics.* Amsterdam: North-Holland Publishing Co.

Herman, G. (1971) Models for Cellular Interaction in Development. *International Journal of Systems Science, 2,* 271-290.

Howland, N. (1962) Structural, Hydraulic and Economic Aspects of Leaf Venation and Shape. In *Biological Prototypes and Synthetic Systems,* E.E. Bernard and M.R. Kare, eds. (New York: Plenum Publishing Corp.)

Keller, E.F. and Segel, L.A. (1970) Initiation of Slime Mold Aggregation Viewed as an Instability. *Journal of Theoretical Biology, 26,* 397-415.

Lindenmayer, A. (1968a) Mathematical Models for Cellular Interactions in Development I. *Journal of Theoretical Biology, 18,* 280-299.

Lindenmayer, A. (1968b) Mathematical Models for Cellular Interactions in Development II. *Journal of Theoretical Biology, 18,* 300-315.

Lindenmayer, A. (1971) Developmental Systems Without Interactions; Their Languages and Grammars. *Journal of Theoretical Biology, 30,* 455-484.

Mostow, G.D. (ed.) (1975) *Mathematical Models for Cell Rearrangement.* (New Haven: Yale University Press)

Onsager, L. (1931) *Physical Review, 37,* 405.

Opatowski, I. (1945) On the Form and Strength of Trees III. *Bulletin of Mathematical Biophysics, 7,* 1-4.

Othmer, H.G. and Scriven, L.E. (1971) Instability and Dynamic Pattern in Cellular Networks. *Journal of Theoretical Biology, 32,* 507-537.

Prigogine, I. (1955) *Introduction to the Thermodynamics of Irreversible Processes.* Springfield: C.C. Thomas, Publishers.

Prigogine, I. and Glansdorff, P. (1971) *Thermodynamic Theory of Structure, Stability and Fluctuation.* New York: John Wiley and Sons, Inc.

Rashevsky, N. (1940) An Approach to the Mathematical Biophysics of Biological Self-Regulation and of Cell Polarity. *Bulletin of Mathematical Biophysics, 2,* 15-25.

Rosen, R. (1967) *Optimality Principles in Biology.* London: Butterworths Publishing, Inc.

Rosen, R. (1968) Turing's Morphogens, Two-Factor Systems and Active Transport. *Bulletin of Mathematical Biophysics, 30,* 493-499.

Rosen, R. (1971) Some Comments on the Concepts of Regulation and Positional Information in Morphogenesis. *International Journal of Systems Science, 3,* 325-340.

Scheraga, H.A. (1969) Calculations of Conformations of Polypeptides from Amino Acid Sequences. In *Symmetry and Function of Biological Systems at the Macromolecular Level,* A. Engström and B. Strandberg, eds. (New York: John Wiley and Sons, Inc.)

Thom, R. (1972) *Stabilité Structurelle et Morphogenese.* New York: Benjamin Company, Inc.

Turing, A.M. (1952) The Chemical Basis of Morphogenesis. *Philosophical Transactions of the Royal Society, B237,* 5-72.

Ulam, S. (1962) On Some Mathematical Problems Connected with Patterns of Growth in Figures. In *Proceedings of the Symposium in Applied Mathematics, Volume 14.* (Providence: American Mathematical Society)

Waddington, C.H. (1957) *The Strategy of the Genes.* New York: Allen and Unwin, Ltd.

Wolpert, L. (1968) The French Flag Problem. In *Towards a Theoretical Biology: Volume 1, Prolegomena,* C.H. Waddington, ed. (Chicago: Aldine-Atherton, Inc.)

Wolpert, L. (1969) Positional Information and Pattern Formation. In *Towards a Theoretical Biology: Volume 3, Drafts,* C.H. Waddington, ed. (Chicago: Aldine-Atherton, Inc.)

LECTURE III

In the preceding lecture we considered a variety of mathematical models of the basic morphogenetic mechanisms. We saw that these mathematical formalisms allowed us to obtain important insights into the morphogenetic capabilities of the elementary mechanisms, and indeed were applicable to wide classes of processes which generate form in the biological world.

In the present lecture, we introduce another fundamental concept related to the generation of form in biology, namely the *genome*. If there is any single characteristic of living systems which sets them apart from technical artifacts or other kinds of inorganic systems, it is the presence of a genetic mechanism, or genotype, which preserves the information describing the biological capabilities of the organism. This genotype, when read out (i.e. translated or interpreted) in a particular set of environmental circumstances, is responsible for all of the biological structures and behaviors exhibited by the organism. The totality of these structures and behaviors comprises the *phenotype* of the organism. There is a delicate and exceedingly complex inter-relationship between genotype and phenotype, since as we noted in Lecture I, selection acts on phenotypes, but is measured in terms of genotypes. It is the genes which replicate and which are transmitted to subsequent generations; the phenotypes, corresponding to the Weissmannian soma, do not replicate, but must be re-created through developmental processes which are themselves under genetic control.

Since the genome controls all aspects of somatic or phenotypic organization, it therefore plays a crucial role in the determination of organic form, and in the processes which generate form. In the present lecture, we shall review what is known about primary genetic mechanisms, and the manner in which these mechanisms are involved in form and form generation.

We begin by reviewing briefly the history of our modern concept of the genome and its activities. Insofar as there was a theory of heredity before Mendel, it was primarily a theory of "blending inheritance" in which hereditary factors were infinitely divisible and infinitely miscible. Mendel, on the other hand, proposed an atomic theory of indivisible hereditary particles, or genes, which pass unchanged to subsequent generations; in this, he seems to have been much influenced by the chemistry of his day, where it had been shown that all the infinite diversity of chemical substances arose out of the combinations of a small number of chemical elements which could be recovered unchanged from any compound into which they had entered. Mendel's work was forgotten until the dependence of Darwin's theory of evolution on hereditary mechanisms was fully recognized; it then became mandatory to clarify the hereditary mechanisms in order to have a clear understanding of natural selection. Given the facts of mitosis, it was reasonable to expect that the genetic units postulated by Mendel were to be found in the nuclei of cells; mitosis, gametogenesis and fertilization had been carefully studied by that time. One of the great syntheses of biology was the identification of the morphological structures called chromosomes with the theoretical "linkage groups" of the early geneticists, and the discovery that the formal constructs required to account for genetic observations, such as inversions, crossing-over, and deletions, all had visible chromosomal counterparts. Indeed, using the giant chromosomes of Drosophila, chromosome maps could be made, indicating that the genes themselves were linearly arranged along the chromosomes.

Another great synthesis came about with the fusion of genetics and biochemistry. It was discovered in the 19th century that biological processes were governed by specific catalysts, or enzymes, which invariably turned out to be proteins. Insofar as the properties of a biological system reflect the properties of the chemical reactions occurring within it, these properties are then determined by the amount and character of the catalysts which the system contains. These catalysts themselves must be under genetic control. The suggestion inevitably grew that, in fact, the genes operated precisely through the determination of the kinds and amounts of enzymes which would be present in a cell at any time. With this hypothesis, a chemical theory of heredity became possible, and attention was focused on the synthesis of proteins as the mode for understanding the precise chemical basis for the genetic control of organic processes.

Until about 1950, the genes themselves were generally thought to be proteins. By that time, however, a succession of experimental findings had cast doubt on this idea, and had implicated instead the nucleic acids in primary genetic processes. For one thing, it was the amount of DNA, and not protein, which precisely doubled in the nuclei of dividing cells. For another thing, DNA absorbed ultraviolet light most strongly at frequencies which were most efficient at inducing mutations. The phenomenon of viral infection, in which only nucleic acids were transferred from the virus to the host cell, provided yet another line of evidence. The matter was essentially settled when Watson and Crick proposed their model for DNA structure, and with it both a mechanism by which the genes could store information pertaining to protein structure, and a mechanism accounting for the replication of the gene itself. From this arose the prevailing "dogma" of molecular biology: DNA → RNA → protein. More specifically, it was argued that the linear arrangement of the constituent monomers in a DNA molecule could *code* for the linear arrangement of constituent monomers in a corresponding polypeptide, and that the synthesis of a specific polypeptide amounted to the *translation* of a genetic message, written in a four-letter alphabet, into a new phenotypic message, written in a twenty-letter alphabet. The code itself was "broken" by Nirenberg and others, and important new insights into mutations and the phylogeny of proteins was obtained thereby. For further details, see Watson (1965).

This picture seemed simple, universal, and conceptually satisfying. But an essential ingredient was still missing. This ingredient pertained to the *control* of genetic activities. For instance, it must be the case that every cell in a multicellular organism possesses the same genome, since these cells constitute a clone. Yet through a process of differentiation, these cells ultimately become very different from each other; there is almost no morphological similarity between a neuron, a liver cell, and an epithelial cell. If the genes do control every aspect of organic structure, how is it possible to account for the phenomena of differentiation? Obviously, we must introduce some mode of differential control for the activities of individual genes in the genome. A mechanism whereby this control could be accomplished without violating the other facts of molecular genetics was suggested by Jacob and Monod (1961) on the basis of their work with bacteria; this was the *operon hypothesis*, which we must now briefly describe.

Bacteria exhibit certain behaviors highly reminiscent of differentiation in higher organisms. Prime among these is the phenomenon of "bacterial adaptation" to different substrates. Thus, bacteria grown in a particular medium, with a particular carbon source, exhibit precisely those enzymes required to metabolize that carbon source. When the medium is changed and a different carbon source provided, the bacteria lose the enzymes they initially exhibited and instead produce a new set, appropriate to the metabolism of the new carbon source. This process is reversible, and allows the bacteria to function adaptively and effectively in a wide variety of environments. Two bacteria of the same clone, grown in two different media, will then be different in enzymatic constitution i.e. differentiated from each other, even though they have the same genome.

On the basis of their experimental work with bacterial adaptation, Jacob and Monod suggested that a bacterial gene is a composite structure, built out of several functional elements. On the one hand, there is a segment of DNA which they called the *structural gene*, which codes for a specific protein. Attached to this is another element, called an *operator*. Elsewhere in the genome is a segment called a *regulator gene,* which produces a small molecule which can combine reversibly with the operator. When the operator is bound by this small molecule, the corresponding structural gene is turned off i.e. cannot be transcribed into messenger RNA nor translated into protein. Thus, the molecule produced by the regulator gene is called a *repressor*. In this picture, then, the "natural" state of a structural gene is to be turned off. However, the repressor itself can combine with other molecules in such a way as to prevent interaction with the operator. When this happens, the repression ceases and the structural gene becomes active. Accordingly, molecules which prevent the action of the repressor, are called *inducers*. A single operator may be associated with several structural genes, and the complex of an operator and the structural genes it controls is called an *operon*.

Since the inducers and repressors for a particular operon may themselves be products of reactions catalyzed by enzymes coded by other genes, the entire bacterial cell can then be regarded as a *network* of interacting operon units. It is clear that such cells can become differentiated from each other, even though they contain the same structural genes. Jacob and Monod explicitly suggested that the phenomenon of differentiation in higher organisms could be understood in the same terms. Thus, attention immediately turns to the properties of networks of such operators, and how the networks may be interpreted in terms of differentiation. It is equally clear that this is a rich area for mathematical analysis,as we will now proceed to indicate.

In reading the original work of Jacob and Monod the first thing one thinks of is that the operon unit which they propose is essentially an all-or-none element; inputs either elicit a response from the element (the inducers) or they suppress this response (the repressors). The output of the element is the product of the reaction catalyzed by the protein and coded by the structural gene of the operon. That is, we have a formal module of the following schematic type.

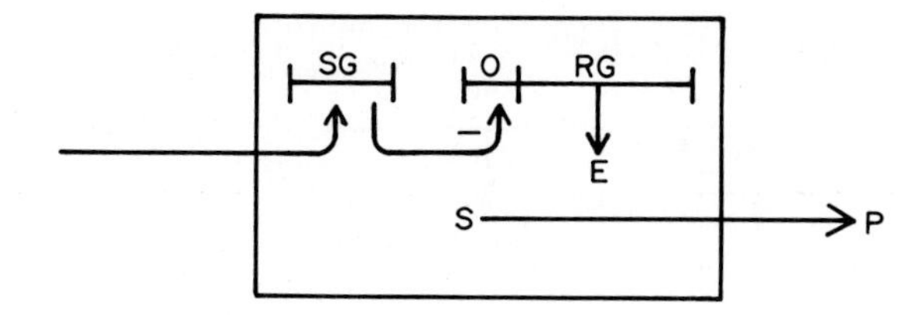

Such a module is reminiscent of other types of excitable elements studied in mathematical biology, especially those which occur in the study of the central nervous system. However, it must be noted that the operon module is a *functional* unit and not an anatomical one; for example, its operation involves the protein-synthesizing machinery of the entire cell.

The similarity between networks of operons and other types of networks of excitable elements is reinforced by consideration of some of the simple networks proposed by Jacob and Monod as examples of how differentiation could occur. Thus, these authors proposed the network

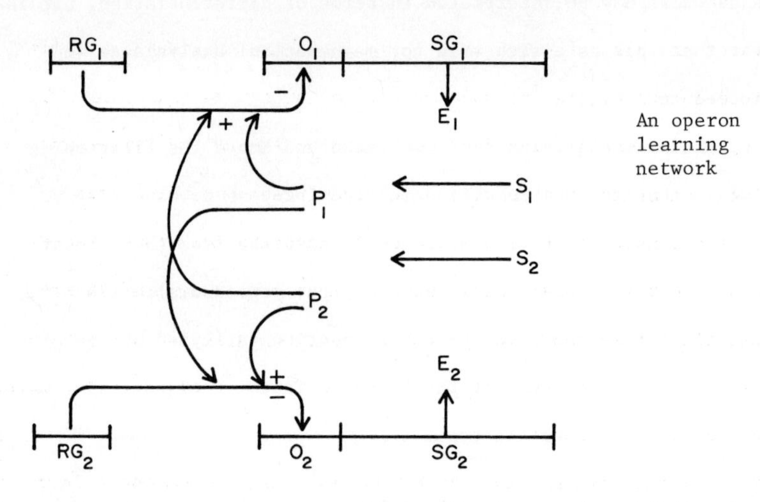

as an example of a two-gene system which will ultimately exhibit only one of the possible "phenotypes". They suggested the network

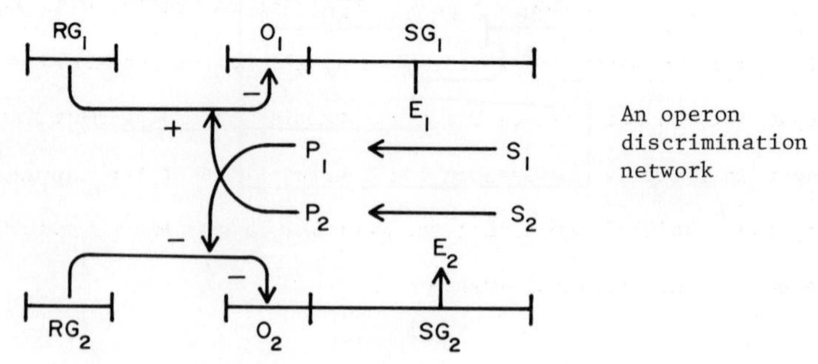

as an example of a two-gene system which could produce one of two possible "phenotypes", depending on environmental circumstances. These networks are substantially identical to those proposed by Landahl and Rashevsky some 25 years earlier (cf. Rashevsky, 1960) for learning and conditioning in neural networks.

These remarks suggest that the entire literature of excitable networks, originally developed for the central nervous system, can be taken over entirely into a study of operon networks. This is indeed the case. We shall now examine some consequences of this fact; later we shall discuss some of the theoretical implications.

There are basically two ways in which we can approach the study of excitable elements and the networks which can be formed from them. On the one hand, we can exploit the fact that the fundamental variables are concentrations, changing as a result of chemical reactions which are governed by rate constants; this enables us to describe the properties of our units and networks in analytical terms by means of systems of first-order differential equations. On the other hand, we can exploit the all-or-none feature of our units and treat our networks in discrete terms, using methods from the theory of automata.

The earliest attempt to extract mathematical consequences from the Jacob and Monod picture utilizing analytic techniques was that of Brian Goodwin (1963). Goodwin's point of departure was the simplest possible operon network, consisting of a single operon repressed by its own output. If x represents the concentration of the primary gene product (messenger RNA) and y represents the concentration of the corresponding protein, then Goodwin argued that the behavior of the operon could be represented by the dynamical system

$$\frac{dx}{dt} = \frac{A}{B + Cy} - D$$

$$\frac{dy}{dt} = Ex - F \qquad \text{(III.1)}$$

This is a system with a single, neutrally stable steady state, surrounded by neutrally stable closed trajectories. Thus, a perturbation of the system away from the steady state will evoke a periodic response in this negative-feedback control system. Goodwin was interested in such periodicities as the basis for cyclic cellular processes which could play the role of biological clocks, and to which general temporal processes in cells could be referred. To show that such oscillations were automatic consequences of the Jacob and Monod picture of genetic regulation would provide an intrinsic basis for understanding temporal processes in general.

Goodwin showed how his results could be extended to certain more complex networks of such units. He then proceeded to imbed these networks in a fluctuating environment which was manifested through the stochastic variability of the parameters appearing in equations (III.1). He argued that such systems could be treated via the formalism of statistical mechanics, in accord with the earlier ideas of Kerner (1957), who applied these techniques to the mathematically similar Volterra equations; the resulting analysis gave biological equivalents for the physical concepts of temperature and free energy which could be interpreted in developmental terms. For further details, we refer the reader to the original works; see also Goel *et al* (1970). Unfortunately, some of the implicit assumptions in (III.1) are biologically implausible e.g. the removal of the reactants x and y from the system at a constant rate, independent of concentration.

When these assumptions are relaxed, the oscillations disappear (cf. Rosen, 1972). However, by introducing a stoichiometric parameter α into the equations to obtain a system of the form

$$\frac{dx}{dt} = \frac{A}{B + Cy^{\alpha}} - Dx$$

$$\frac{dy}{dt} = Ex - Fy \quad ,$$

it is possible to restore oscillatory behavior, in the form of limit cycles. We shall return to systems of this form, in another context, in Lecture IV; they are presently the subject of much analysis.

In general, the study of complex networks of operon units described in terms of ordinary differential equations leads to analytically intractable systems. Indeed, this was one of the reasons why this kind of approach to the central nervous system was largely abandoned, in favor of more tractable automata-theoretic approaches. However, quite a number of authors have proceeded with elaborate computer simulations of such systems. The most extensive of these is due to Heinmets (1966, 1970); he has been most concerned with the response of such systems to parameter variations (such as changes in rate constants), and to the interpretation of these responses in terms of differentiation and carcinogenesis.

As we have noted, the analysis of complex networks of excitable elements is generally refractory to analytical approaches. In 1942, McCulloch and Pitts suggested that a more effective mode for studying the properties of such networks was in terms of digital, discrete properties embodied in their formal neurons and the networks composed of them. Almost immediately upon publication of the operon concept, it was recognized by Sugita (1961) and others that the operon units could be similarly treated; since then, much has been written on automata-theoretic approaches to genetic control, which we shall now briefly review (for a more extensive review, see Arbib, 1972). Most of this literature has extensive parallels with earlier work in neural nets, as is, of course, not surprising.

It was recognized very early that the theory of automata (including finite-state machines, and the generalization to Turing machines) can be developed in several distinct ways. On the one hand, we can take a modular approach, regarding an automaton as a network of elementary units, connected in a particular fashion, as in the original McCulloch and Pitts work. Or we can take a more abstract approach, regarding an automaton as a device which can be in one of a finite set S of states at any (discrete) instant of time, and which can receive at that instant any one of a finite set A of inputs. The behavior of the automaton is then embodied in a next-state mapping $\delta: S \times A \to S$, which corresponds to the *program* of the automaton, and in a related output map $\lambda: S \times A \to B$, where B is a finite set of *outputs*. Every neural net gives rise to a description of this type; conversely, any automaton can be *realized* (generally in many ways) as a neural net.

It is possible to generalize this last definition of an automaton from a device which processes individual inputs at specific instants to a device which processes *strings* of inputs, or words in the free semigroup $A\#$ generated by the input set A. Thus, we can regard an automaton as a word processing device, which converts input words into output words. This fact serves, on the one hand, to convert the theory of automata into a branch of algebra, and on the other hand it stimulates suggestive parallels to the biological notion of coding which we have discussed above. Quite different, though related, automata-theoretic approaches to genetic control have been devloped by various authors, by exploiting either the modular network view or the algebraic view of automata as processors of strings of symbols.

Representative of the modular viewpoint is the work of Sugita (1961),or Apter (1966), or Kauffman (1970). Kauffman's work, for instance, is an extension of studies on random networks going back to Shimbel and Rapoport (1948). Kauffman argues that since the details of interconnection of a genetic network are too complex ever to be discovered in detail we may as well assume at the outset that the interconnections are random. Moreover, since we cannot generally discover which of the possible logical functions is actually computed by a particular module, we may as well assume that these functions are randomly distributed as well. He imposes only the constraint that each module in the network receives inputs from only two other modules. On the basis of extensive computer simulations, he suggests that such networks will have, on the average, the interesting property that, if started in some initial state and left to function autonomously in a constant environment, the network will oscillate (this much is trivial) through a cycle of states which is very small compared to the total number of states (he suggests that, on average, the length of such a cycle will be $\sqrt{N}$, where N is the number of elements in the network). Kauffmann then goes on to identify such state cycles with states of differentiation, and draws a variety of conclusions regarding the response of such networks to perturbations. He suggests that this behavior is destroyed if more than two inputs per module are allowed.

Apter's work (1966) is of a rather different character, and is concerned with actually constructing networks which exhibit biologically interesting behavior (as did the simple networks proposed by Jacob and Monod). This was the approach originally suggested by McCulloch and Pitts who pointed out that neural networks (and in general, finite automata) were *universal* in the sense that, given any sufficiently well-formulated behavior to be exhibited by a biological system, we can construct an automaton which will exhibit precisely this behavior. Thus, if we wish to account for any particular pattern of differentiation, we can construct explicit networks of operons which will exhibit precisely this pattern. Generally, of course, there will be many networks which will satisfy the given conditions, but we may hope, by suitable experimental procedure, to determine which of them is actually embodied in the differentiating system. Crucial to this method of proceeding, of course, is the assumption that a differentiating system *is* directly representable as an operon network, and therefore the system must be a realization of one of the family of networks which exhibit the behavior in question. A critique of this point of view, presented from the standpoint of the central nervous system, appears in Rosen (1969).

The idea that we can construct a system to imitate the operation of any given system is central to Wiener's idea of replication and self-reproduction (Wiener, 1961). Wiener argues that it is possible to attach a "white box", capable of producing any behavior (e.g. capable of generating any linear combination of a complete set of functions), to a "black box" in such a way that the parameters of the white box are incrementally adjusted until the behavior of the "black box" is matched. Similar ideas are also implicit in the learning models of Grossberg (1974), which are likewise developed from a modular point of view. It should be noted that problems of biological adaptation, and ultimately of evolutionary development, can be approached in this way, although we cannot discuss this aspect here.

The idea of string processing, implicit in the algebraic view of automata theory, represents a somewhat different kind of approach. We have already noted the fact that the biologically important macromolecules, such as protein and DNA, are linear sequences, or words, built out of a relatively small number of monomeric units, which constitutes an alphabet. We have used the words "coding", "transcription", and "translation" in referring to their biological activities. This fact has led a number of authors to look upon biological systems as essentially string processors which are immediately amenable to automata-theoretic treatment independent of any idea of modular genetic networks. Thus, Stahl *et al* (1964) were led to model a biological cell as a Turing machine, on whose input tape was inscribed the genome of the cell, as well as all the precursors and metabolites required for the construction of the important macromolecules,and an initial complement of enzymes to process these materials. This "algorithmic cell" embodies a faithful translation of the idea that a biological cell is simply a processor of strings of symbols. Stahl's approach can be regarded as a digital analog of the simulations of Heinmets mentioned earlier. It should be noted explicitly, however, that in this approach the genome of the cell is not embodied in its program (i.e. the state-transition function) but is part of the data. The program itself depends, in these models, on the interactions specified between the various cellular constituents.

A rather more elaborate view of the organism as an automaton arises from extensions of von Neumann's work on self-reproducing automata (cf. Burks, 1966). This work is of an exceedingly abstract character, and is difficult to interpret in concrete biological terms. Basically, von Neumann's universe of discourse is a space divided into cells, much as in our models for sorting-out described in the previous lecture. In each of these cells reposes a copy of an automaton A, capable of one of a number of internal states. These automata can communicate with their nearest neighbors, and change state in accordance with outputs produced by the neighbors. Let one of the states be designated by the symbol Q (quiescent), and let us suppose that initially all the automata are in the state Q except for some finite configuration. That configuration is *self-reproducing* if, at some later time, the original non-quiescent configuration, together with a copy of it elsewhere in the tessellation, are non-quiescent, with all other cells quiescent. It is clear that we can speak in such circumstances, not only of "self-reproduction", but also of "growth", "development", and "differentiation".

It is clear that these tessellation models go far beyond the properties of simple operon networks; networks do not "reproduce themselves" but merely control the state of their own activity as a function of their interconnection and of environmental stimuli. It is true that we can interpret some of these activities as being concerned with replication, but this involves an *ad hoc* aspect not connected with the formal properties of the networks. I believe it is fair to say that the properties of tessellation automata have no immediate biological counterparts beyond the fact that the same words, such as "growth", can be applied in both simulations.

REFERENCES

Apter, M. (1966) *Cybernetics and Development.* New York: Pergamon Press, Inc.

Arbib, M.A. (1972) Automata Theory in the Context of Theoretical Embryology. In *Foundations of Mathematical Biology, Volume 2,* R. Rosen, ed. (New York: Academic Press, Inc.)

Burks, A. (1966) *Theory of Self-Reproducing Automata.* Urbana: University of Illinois Press.

Goel, N.S., Naitra, S.C. and Montroll, E.W. (1971) On the Volterra and Other Nonlinear Models of Interacting Populations. *Review of Modern Physics, 43,* 231-276.

Goodwin, B. (1963) *Temporal Organization in Cells.* New York: Academic Press, Inc.

Grossberg, S. (1974) Classical and Instrumental Learning by Neural Networks. In *Progress in Theoretical Biology, Volume 3,* R. Rosen and F.M. Snell, eds. (New York: Academic Press, Inc.)

Heinmets, F. (1966) *Analysis of Normal and Abnormal Cell Growth.* New York: Plenum Publishing Corp.

Heinmets, F. (1970) *Quantitative Cellular Biology.* New York: Marcel Dekker, Inc.

Jacob, F. and Monod, J. (1961) On the Regulation of Gene Activity. *Cold Spring Harbor Symposia on Quantitative Biology, XXVI,* 193-211.

Kerner, E. (1957) A Statistical Mechanics of Interacting Biological Species. *Bulletin of Mathematical Biophysics, 19,* 121-146.

Kauffman, S. (1970) Behavior of Randomly Constructed Genetic Nets. In *Towards a Theoretical Biology: Volume 3, Drafts,* C.H. Waddington, ed. (Chicago: Aldine-Atherton, Inc.)

Monod, J. and Jacob, F. (1961) General Conclusions: Teleonomic Mechanisms in Cellular Metabolism. Growth and Differentiation. *Cold Spring Harbor Symposia on Quantitative Biology, XXVI,* 389-401.

Rashevsky, N. (1960) *Mathematical Biophysics, Volume II.* New York: Dover Publications, Inc.

Rosen, R. (1969) Hierarchical Organization in Automata-Theoretic Models of the Central Nervous System. In *Symposium on Information Processing in the Nervous System,* K.N. Leibovic, ed. (New York: Springer-Verlag New York, Inc.)

Rosen, R. (1972) *Dynamical System Theory in Biology*. New York: John Wiley and Sons, Inc.

Shimbel, A. and Rapoport, A. (1948) Steady States in Random Nets. *Bulletin of Mathematical Biophysics, 10*, 211-220.

Stahl, W.R., Coffin, R.W. and Coheen, H.M. (1964) Simulation of Biological Cells by Systems Composed of String-Processing Automata. *Proceedings of the AFIPS Spring Joint Computer Conference*. Baltimore: Spartan Books, Inc.

Sugita, M. (1961) Functional Analysis of Chemical Systems in vitro Using a Logical Circuit Equivalent. *Journal of Theoretical Biology, 1*, 413-430.

Watson, J.D. (1965) *Molecular Biology of the Gene*. New York: Benjamin Company, Inc.

Wiener, N. (1961) *Cybernetics, 2nd Edition*. Cambridge: M.I.T. Press.

LECTURE IV

In the previous lecture, we have briefly reviewed what is known about the direct genetic control of biological processes, and some of the attempts to model these processes mathematically. As we have seen, the current conception is that genes act directly by specifying the synthesis of particular functional proteins; these in turn govern the nature and rate of all processes occurring within any particular cell. Thus, present views of primary genetic mechanisms are primarily biochemical.

On the other hand, it has long been known that, in multicellular organisms, numerous gross alterations of morphology in the whole organism behave like single gene mutations. The original pea mutation studies by Mendel involved the stature of the plant, the color of the cotyledon, and the form of the seed coat. Drosophila genetics developed primarily through mutations which modified the form of the organism bearing them; literally thousands of examples may be found in the classical texts on genetics. Indeed, it should be recognized that the biochemical view of the gene is a relatively late development; the early geneticists all tended to view the action of genes in terms of their morphological and morphogenetic consequences.

There are thus several fundamental gaps between the biochemical processes, known to be under direct genetic control, and the morphological and morphogenetic processes we wish to understand in terms of genetic activity. For one thing, a genome directly controls the biochemical events occurring within the cell in which the genome lies; morphogenesis typically involves gross features of large families of cells and their products. For another, let us note that all of the morphogenetic mechanisms which we discussed extensively in Lecture II are *genome-independent;* the word "gene" was never mentioned, and indeed, the best concrete examples of these mechanisms arise in non-biological systems, where the concept of a genome seems quite meaningless from the outset.

How are we to integrate the idea that the genome of an organism is responsible for its morphogenesis with the two other basic ideas we have developed: (a) that the genes are directly responsible for biochemical events, and (b) that the basic morphogenetic mechanisms do not require a genome? It is with this question that the present lecture, and the next, will be concerned.

We will begin with some prefatory remarks. One of the oldest controversies in biology, which has its roots in the study of developmental processes, is that of *preformation* versus *epigenesis*. The preformationist view held that all the structures typical of the adult organism were already present in the zygote, and only a passive growth process was necessary to reveal them. That is, developmental processes required no increase of initial complexity or the emergence of new properties in passing from zygote to adult; all of the ultimate structures of the adult must pre-exist in the zygote. The epigenetic view, on the other hand, held that development did in fact begin with a poorly organized zygote, and that there is a real increase in complexity and organization arising from a dynamical process. Thus, in the epigenetic view, there was no pre-existing structural basis for the final adult organism in the zygote, but merely an appropriate set of "initial conditions" from which the adult would inevitably emerge. Epigenesis is thus a *creative* process; preformation is not.

The idea of epigenesis, in a more modern guise, provides a framework through which we may approach the fundamental question posed above. The idea is, roughly, that the genome serves to specify an appropriate set of "initial conditions" for subsequent dynamical processes; it is these subsequent dynamical processes which are the putative objects of study in morphogenesis, but the individual trajectories of these processes, which correspond to specific modes of morphogenesis, cannot be specified without the concordant specification of initial conditions, and these reside in the genes. In the remainder of these lectures, we shall try to put some flesh on these intuitive notions, and see how far they may lead.

Let us start at the most elementary level. As we have seen, the primary activity of a structural gene is to "code" for the production of a specific polypeptide. More specifically, the linear sequence of nucleotides in a strand of DNA is "translated" into a corresponding linear sequence of amino acids in a polypeptide. This linear sequence is often called the *primary structure* of the polypeptide, and it is the primary structure which is "coded for" by the structural gene. However, as we saw earlier, the polypeptide must then spontaneously, through a dynamic process, *fold up* into a specific three-dimensional conformation, or *tertiary structure*, which also confers upon it its specific functional properties.

It is a fundamental dogma of molecular biology that the tertiary structure (i.e. the biologically active conformation) is uniquely determined by the primary structure, coded for by the genes. However, intervening between a newly synthesized polypeptide of given primary structure, and the biologically active conformation of that polypeptide, is a dynamical process of folding. This folding process is not directly coded for in the genome, but must occur spontaneously, through a process of self-assembly or self-organization. We have seen above how we may understand such a process in terms of the attainment of a minimal free-energy state in a population of (coupled) motile units with differential affinities for themselves and their environment. But the main point is that the folding process arises from properties automatically possessed by real sequences of amino acids (i.e. their differential affinities for one another and for the ambient medium), but not directly coded for in the genome; the genome only codes for the primary structure of the sequences. The folding itself, and the active conformation it generates, is then an archetypal example of an epigenetic process. The structural gene acts to set initial conditions for a dynamics not itself coded by the genome; the activity of this process depends on properties of the constituent amino acids *other than* their linear relation to each other (which is precisely what the genome does code for).

Other examples of self-organization and self-assembly at the molecular level can also be regarded as epigenetic in this sense. The spontaneous formation of the typical cylindrical tobacco-mosaic virus coat from its protein subunits depends entirely upon the tertiary structures of those subunits. Their sites of association with one another are not directly coded for in the viral genome, but arise as a result of dynamical folding processes for which the primary sequence specifies an appropriate set of initial conditions. The association of these folded subunits to form the viral coat in turn arises out of another kind of dynamic process, for which the tertiary structure of the subunits sets the initial conditions. Here, too, the basic requirements of an epigenetic process are present: the ultimate form is generated by a dynamical process of a very general kind, not itself "coded for" in any sense, and exploiting properties of the individual reactants which are *different from* those which are directly "coded for". The "coded" properties then serve to set initial conditions for the dynamical process which ultimately generates the final form.

This *leitmotif* of epigenesis pertains, not only to structures formed by self-assembly, but also to dynamical processes arising from the interactions of gene-coded products. Let us consider one important example, which has been the object of much study - the epigenetic control of enzyme-catalyzed reaction sequences in cells.

We have already described some of the phenomena of adaptation in bacteria, as they relate to the synthesis of metabolically active enzymes; this is embodied in the operon concept. However, there are regulatory phenomena connected with adaptation which occur directly at the enzyme level, and which have to do with manipulation of the rate constants of these enzymes by metabolites. These phenomena are typically found in synthetic pathways; a good example is that involved in the synthesis of the amino acid tryptophane from simpler precursors. It is necessary for all protein-synthesizing cells, such as bacteria, to maintain pools of amino acids for incorporation into their protein. This synthesis typically takes place in a sequence of steps, each catalyzed by a particular enzyme:

$$A_0 \xrightarrow[\text{source}]{\text{external}} A_1 \xrightarrow{E_1} A_2 \xrightarrow{E_2} A_3 \xrightarrow{E_3} \cdots \xrightarrow{E_{n-1}} A_n$$

It was found experimentally that, if the end-product of this reaction sequence was directly provided to bacteria, the entire synthetic pathway leading to that end-product was inhibited. This inhibition would occur even if there were irreversible steps in the sequence, or if the sequence was interrupted by the removal of one of the enzymes. It was discovered that the first enzyme in the specific pathway leading to the end-product was specifically and non-competitively inhibited by the end-product. In fact, a separate "control" site, distinct from the active site, was found on this first enzyme; when bound with end-product this control site completely inhibited the enzyme. Thus, there are at least two levels of control of enzymatic processes in cells; one of these acts directly at the genetic level and governs the synthesis of enzyme, while the other acts epigenetically at the enzyme level, to control the rate of enzymatic reaction.

A feedback-controlled reaction system like that outlined above is a direct generalization of the mechanism we have already considered for feedback repression. Such a system may be modelled by a system of rate equations of the form

$$\frac{d[A_1]}{dt} = \frac{d_0[A_0]}{1+K[An]^{\rho}} - k_1[A_1],$$

$$\frac{d[A_i]}{dt} = k_{i-1}[A_{i-1}] - k_i[A_i], \quad i = 2,\dots,n.$$

Much interest has been focused on the possibilities of oscillations in such systems. (See Morales and McKay, 1967; Higgins, 1967; Walter, 1969).

There are many other regulatory properties inherent in enzymatic reaction sequences besides end-product inhibition of the type described above. Enzymes may be activated (i.e. their rate constants may be enhanced) as well as inhibited by interactions with other metabolites. Activation was found to be involved in an enzyme system observed experimentally to produce oscillations (Chance *et al*, 1967). Moreover, just as we can have networks of interacting operons connected by induction and repression, so, too, at the epigenetic level we can have networks of interacting enzyme systems connected by activation and inhibition. Such is the case, for example, between the enzymatic systems responsible for the production of high-energy phosphate bonds; one of these systems requires oxygen (the familiar Krebs cycle) while the other is anaerobic (the Emden-Myerhof pathway). Moreover, free oxygen inhibits the anaerobic system (this is called the Pasteur effect) while the Krebs cycle is inhibited by excess carbohydrate (the Crabtree effect or glucose effect). For further details, see Chance *et al* (1960).

It is clear that all of these adaptive phenomena are epigenetic in character. An oscillation arising in a feedback-controlled, enzyme-catalyzed reaction sequence is not in any sense directly coded for by the genome, but arises from dynamics of a general kind for which the genome ultimately sets the initial conditions.

Beyond this relatively concrete biochemical level, discussions of epigenetic mechanisms become much less detailed. The earlier literature on developmental biology is filled with arguments concerning "morphogenetic fields"; this terminology was introduced into biology originally by analogy with phenomena such as the morphogenetic properties exhibited by iron filings in magnetic fields (cf. the interesting discussion of the origin of field concept in Needham (1950, p. 127 ff)). Later, when a link with chemistry had been made through the discovery of the organizers, and with the development of notions of competence and evocation (cf. Lecture I), these fields assumed the character of chemical gradients of evocators, inducers and organizers acting on domains of differentially competent materials (Dalcq, 1938; Child, 1941).

It is clear that a discussion of development processes in these terms is necessarily much less detailed than the manner in which we can treat processes such as the folding of proteins or the properties of sequences of catalyzed reactions. Indeed, discussions of epigenetic phenomena are very theoretical in character, although not (at least initially) of a mathematical nature, even though they are couched in a language directly related to empirical observation. Nevertheless, the language of gradients and fields, of inductions and competences, is tantalizingly close to the language of operons, repressors and anti-repressors, of catalyzed reactions and metabolism, so that the next step seemed, inevitably, to be that of obtaining an integrated theory of biochemistry and development which would place the phenomena of morphogenesis in a chemical context, just as primary genetic processes had been placed in such a context.

On the theoretical side, this inevitable step has resulted precisely in the morphogenetic mechanisms which we discussed in Lecture II. These are all dynamical mechanisms couched in biochemical language; we have spoken of free energies, of cells, of chemical reactions and diffusions, of morphogens, and of the transfer of positional information through the propagation of chemical oscillations. The reason for studying these mathematical formalisms is that they alone can exhibit for us the morphogenetic capabilities inherent in such systems and tell us what can be accomplished in systems whose elements exhibit the biochemically plausible properties revealed by empirical studies. Conceptually, it will be noted that all of the morphogenetic mechanisms mentioned in Lecture II deal with *epigenetic* mechanisms; they tell us what the dynamical consequences of a particular set of biochemical hypotheses will be; they determine individual trajectories once appropriate initial conditions, superimposed on those biochemical hypotheses, have been specified. The *generation* of the interactive capacities on which a particular morphogenetic mechanism is based and the setting of the appropriate initial conditions must be specified *a priori,* which means that they lie *outside the scope of the morphogenetic mechanism* itself. These capacities and initial conditions cannot, in principle, be predicted from this kind of formalism, which can only describe their consequences. Intuitively, as we have argued, they must come from the genome, which thus sits as a *deus ex machina* outside the morphogenetic process in all of the mechanisms we have discussed.

Furthermore, despite the use of a language which seems similar to the language of empirical biochemical events, and is therefore capable of experimental interpretation, it has proved most difficult to obtain empirical predictions from the morphogenetic mechanisms we have described, and corresponding experimental verification or falsification. Since this fact has often been used as an argument against the validity or utility of mathematical theory in biology, it is necessary to investigate the matter in some detail. Other aspects of the same question will be discussed in Lecture V. We shall now concern ourselves with the idea that the morphogenetic mechanisms described in Lecture II are not individual models of morphogenesis but rather comprise large classes of such models, related through some common principle which all exhibit.

In our previous biological discussions of genetic processes we have implicitly stressed the structural *differences* between organisms of different kinds. Different genes code for different primary structures in proteins; these in turn fold in such a way as to generate distinct conformations, with diverse specificities and control properties. The products of the reactions catalyzed by these diverse proteins interact differently with regulator genes, giving rise to different connectivities in the operon networks which represent the integrated activity of the genome. This is in accord with the basic ideas of molecular biology, which seeks to explain the differences between organisms in terms of differences of structure; these, in turn, must ultimately arise from differences in the molecular populations comprising those organisms. Since organisms can be so diverse at the molecular level, it would appear that a new theory, based on *idiosyncracies* (rather than commonalities) of molecular structure, would be required to understand the behavior of every kind of organism.

On the other hand, we are better acquainted with the phenomena of the biosphere in terms of commonalities than in terms of idiosyncracies. Indeed, biological science is founded primarily on our intuitive notion of what constitutes an organism, and this intuition in turn rests upon homologies of behavioral activities displayed by organisms of endlessly diverse physical structures. Every attempt (and there have been many) to specify the class of organisms within the class of all physical systems by using structural or physical characteristics has failed. Most of these attempts utilize structural criteria, which we know play no part in our intuitive recognition of what constitutes an organism (e.g. whether or not DNA is present); moreover, these structural characterizations fail, not because they cannot provide a decision in every case, but because these decisions do not accord with our fundamental, intuitive ideas concerning what constitutes an organism. For example, since we do not intuitively regard a flame or a crystal as alive, we reject those physical characterizations which include such systems among the organic ones; likewise, since we intuitively consider a bacterial spore to be alive, we reject those characterizations which place these systems among the inorganic ones. This judging of the structural descriptions of organic processes by our intuitive idea of what constitutes an organism is an example of the inter-relationship of multiple descriptions which, in another guise, and owing to the influence of Rene Thom, has come to play a significant role in morphogenetic studies; we will consider this matter in Lecture V.

If it is true that the commonalities between organisms comprise the basic subject-matter of biology, then we must come to terms with the fact that such commonalities can be exhibited by systems with very diverse molecular structures. Proteins with very diverse primary structures can fold so as to generate common catalytic specificities. Organisms with very diverse genetic characteristics, in structural terms, can nevertheless exhibit essentially the same morphogenetic behavior. Granted that different organisms are structurally diverse, how can we explain their essential homologies in terms which stress their diversities?

If we look again at the various epigenetic mechanisms described in Lecture II in the light of what has just been said, we will notice some interesting features. Take the Rashevsky-Turing mechanism as a case in point. This mechanism does not pertain to a single system, but rather to *any* system possessing a dynamics in which the homogeneous states are unstable. Systems described by such dynamics can be endlessly diverse, both structurally and mathematically. If we identify the process of morphogenesis with departures from homogeneity, then all of these systems automatically exhibit morphogenesis.

Thus, the Rashevsky-Turing mechanism does not describe a single developmental system but rather serves to potentially describe an enormous class of such systems. Moreover, it describes them, not in terms of the diversities of the dynamics they manifest, but rather in terms of a common feature, the instability of the homogeneous states. The same is true for all of the other epigenetic mechanisms described in Lecture II, and for the theoretical considerations presented in Lecture III and in the present lecture.

A morphogenetic mechanism, such as that manifested in the Rashevsky-Turing system, is what we have proposed (Rosen, 1972) to call a *dynamical metaphor* for morphogenesis. In these terms, a metaphor is a class of diverse systems which exhibit homologous behaviors and manifest a common principle. Thus, a metaphor differs profoundly from what is ordinarily called a *model* of the behavior of an individual real system. A model typically arises from idiosyncracies of (physico-chemical) structure which can be represented mathematically and which correspond, more or less, to observable properties of the real system.

In conceptual terms, then, a dynamical metaphor is a *class of possible models* for a kind of behavior exhibited by a specific real system. To say, then, that the Rashevsky-Turing ideas provide an *explanation* for any specific differentiation phenomenon in biology is to say, roughly, that this specific instance of differentiation arises because the system in question possesses a dynamics which renders its homogeneous states unstable. Already, on this basis, we can acquire insights into the properties of the system in terms of the properties which any dynamical system of this character must evidence. But we are in fact saying much more than this; we are saying, in effect, that *a model appropriate to describe the dynamics of the real system of interest is already a member of the Rashevsky-Turing metaphor.* A full understanding of that dynamics is then to be found by extracting that model from the class of models which constitutes the metaphor.

As long as we remain at the metaphoric level, there is a sharp limit to the precision of a description of any instance of development exhibited by a real system. We are restricted to commonalities, to those properties which are manifested by all the models in the metaphor. The properties of an individual model which rest upon its specific characteristics i.e. upon its idiosyncracies, cannot be extracted at this level of analysis. We believe that it is this fact, more than any other, which has made the results of metaphorical reasoning, particularly in morphogenesis, so difficult to integrate into experimental biology. For, as we have noted, biological experiments on individual systems have concentrated on features which make these systems different from each other rather than on features which they exhibit in common.

The situation we have just described is not different from our experience in mathematics. We can characterize classes of systems in mathematics by virtue of their global properties; for example, we characterize the class of groups among all algebraic systems by virtue of the fact that the group axioms are satisfied. That the group axioms are satisfied does not devolve at all on the nature of the elements of the group, nor upon the nature of the binary operation; we cannot find out whether a system is a group or not by considering these idiosyncracies. It is precisely this fact which gives group theory its extraordinary generality. However, there is not too much of great interest which can be proved directly from the group axioms. There are simply too many kinds of objects which satisfy the group axioms to allow one to make strong statements which are always true. But by requiring additional properties, such as finiteness or commutativity, we characterize a class of groups still broad enough to be interesting, but narrow enough to permit relatively strong results to be obtained.

Having come this far, there are two further observations which must be made; one of these pertains to the abstract level of the dynamical metaphor and the individual models which comprise it, while the other pertains to the *realization* of these individual models.

The first observation is the following: if it is true that a dynamical metaphor is a class of systems sharing a common organizational property, and if it is also true that the differences between the individual systems arise from idiosyncratic characteristics not directly bearing on the commonalities which define the metaphor, then we should be able to *compare* the different systems in the metaphor in terms of the commonalities. The specific vehicle whereby this is done in mathematics is the set of *structure-preserving* mappings carrying one element in the metaphor into another. The study of such sets of structure-preserving mappings in abstract terms has been called *relational biology* (Rashevsky, 1954; Rosen, 1972). On a more concrete level, structure-preserving mappings are precisely what are involved in the comparative theories of form (such as the D'Arcy Thompson Theory of Transformations) which were mentioned in Lecture I. These comparative theories are thus implicit in, and corollaries of, a metaphorical approach to biological regularities, but they are foreign to the *ad hoc* treatment of biological idiosyncracies. These theories embody the idea which underlies our basic intuition regarding the recognition of biological phenomena, the idea that all organisms are, in an important sense, *dynamical models* of each other. This idea, which dominated theoretical biology until 1935, and which still underlies the use of rats and other organisms to illuminate man's physiology and psychology, has been obscured by the emphasis on structural idiosyncracies since that time. Nevertheless, this idea deals with the crucial problems of the biosphere; it also arises naturally in present-day investigations of morphogenesis, and some of the earlier papers on these matters still make provocative reading (cf. Lambert and Teissier, 1927; Needham, 1934).

To close the gap between what we learn from the formulation of a dynamical metaphor, and what we learn from the detailed empirical investigations of individual systems, we must find some way of extracting one, or a small number, of individual systems from the metaphor. The only way to do this is in terms of what we have called idiosyncracies, properties which distinguish one element of the metaphor from another. Such a system may then be characterized in far greater detail, and its properties compared to those of an empirical system we may wish to understand. An example of this is provided by the Keller-Segel approach to slime-mold aggregation, mentioned in Lecture II, which is a member of the Rashevsky-Turing metaphor for differentiation. Formally, how does one go about extracting single elements from classes of otherwise equivalent systems? Obviously, the way to do this is to add constraints, pertaining to the idiosyncracies of these systems, which may be satisfied by some elements of the metaphor and not by others. A prototypic example of such a constraint is universally used in the theory of *optimal processes*. In this theory, a *cost function* is superimposed upon the class of systems; those systems belonging to the class which also minimize (or maximize) the cost function are then extracted. Such a process is akin to what nature itself does in extracting forms through natural selection; here the cost function is called *fitness*. Aspects of this means of selecting specific systems which minimize some cost function from a class of otherwise equivalent systems will be discussed by Dr. Bremermann in his lectures, and related matters are discussed in Rosen (1967). There is much conceptual work to be done (cf. the discussion of multiple descriptions in Lecture V) before this line of reasoning can be applied to extract individual models from the Rashevsky-Turing metaphor, for example, for direct comparison with experiment. Nevertheless, it is felt that this strategy will be more effective than the present alternative i.e. the generation of ad hoc models from structural idiosyncracies, which is still the prevailing mode in experimental biology.

To conclude, we must add a word about *realizations*. We have said that a dynamical metaphor is a class of individual mathematical systems, each of which is a presumptive model for a morphogenetic process. But a model is still an abstract mathematical structure. In order to relate a model to an empirical situation, the formal elements of the model must be identified with specific properties of the real system. Such an identification constitutes a *realization* of the model. Now it may happen that the same model can be realized in many different contexts. For instance, we saw in Lecture III how an abstract network of formal neurons might be realized in a non-neural context, in terms of networks of operons. When this happens i.e. when two different systems S_1, S_2 constitute realizations of the same mathematical formal structure M, then S_1 and S_2 become *models of each other*. This usage of the term "model" (which is badly overworked in the sciences) is distinct from that utilized in the previous assertion that "all organisms are models of each other", but the two are closely related.

The former usage dealt with situations in which a common metaphor applies to different models; the present usage considers situations in which a common model applies to different realizations.

As we shall see in Lecture V, all models arise through processes of abstraction. Since these are, in effect, processes which isolate subsystems from a given system, we may say that two real systems are exact models of each other when they contain isomorphic subsystems which can be described by the same abstract mathematical formalism. This fact provides the basis for the use of "model systems" in experimental biology, the use of thin inorganic films as "model membranes", the use of inorganic catalysts as "model enzymes", the use of switches of various types as "neuromimes", etc. The prevalent, fruitful use of such systems represents further support for the assertion that idiosyncratic structural details do not provide a proper basis for the understanding of function and organization; the ease with which the metaphoric language we have developed accommodates these situations is another indication of its utility.

Furthermore, the sharing of common subsystems, which allows real systems to "model" each other in this sense, introduces a new basis for a taxonomy of systems and enables us to recognize systems as related even though they are structurally very different. This, in turn, provides a new basis for unifying apparently unrelated fields of biology. It is commonly supposed that a unified theory of biology can only arise through reductionistic approaches, which dissect all biological phenomena into a set of common structural elements. However, we have seen that such a structural approach does not readily incorporate homologies of behavior and organization, which in many ways are the basic subject-matter of biology. On the other hand, if we relate systems through their behavioral homologies, as we do in the metaphorical language described above, then the structural elements do not enter into the characterization of these systems at a fundamental level. The kind of unification arising in this way may be extremely powerful, as it has been in physics (cf. the mechano-optical analogy of Hamilton, which unified optics and mechanics through common formal minimization principles, and led directly to the Schrödinger equation), without involving a "reduction" in any sense.

The problems of relating systems through the extraction of common dynamical subsystems, rather than common structural elements, lead naturally to the manner in which different modes of description of the same system may be compared and integrated with each other. These same questions, motivated somewhat differently, will be considered in Lecture V.

REFERENCES

Chance, B., Garfinkel, D., Higgins, J.J. and Hess, B. (1960) Metabolic Control Mechanisms V: A Solution for the Equations Representing the Interaction Between Glycolysis and Respiration in Ascites Tumor Cells. *Journal of Biological Chemistry, 235,* 2426-2430.

Chance, B., Pye, K. and Higgins, J.J. (1967) Waveform Generation by Enzymatic Oscillators. *I.E.E.E. Spectrum, 4,* 79-86.

Child, C.M. (1941) *Patterns and Problems of Development.* Chicago: University of Chicago Press.

Dalcq, A.M. (1938) *Form and Causality in Early Development.* Cambridge: Cambridge University Press.

Higgins, J.J. (1967) The Theory of Oscillating Reactions. *Industrial and Engineering Chemistry, 59,* 18-62.

Lambert, R. and Teissier, G. (1927) Theorie de la Similitude Biologique. *Annales de Physiologie et de Physicochemie Biologique, 2,* 212-246.

Morales, N. and McKay, D. (1967) Biochemical Oscillations in Controlled Systems. *Biophysical Journal, 7,* 621-625.

Needham, J. (1934) Chemical Meterogony and the Ground-Plan of Animal Growth. *Biological Reviews, 9,* 79-109.

Needham, J. (1950) *Biochemistry and Morphogenesis.* Cambridge: Cambridge University Press.

Rashevsky, N. (1954) Topology and Life. *Bulletin of Mathematical Biophysics, 16,* 317-348.

Rosen, R. (1967) *Optimality Principles in Biology.* London: Butterworths Publishing, Inc.

Rosen, R. (1972) Some Relational Cell Models: The Metabolism-Repair Systems. In *Foundations of Mathematical Biology, Volume 2,* R. Rosen, ed. (New York: Academic Press, Inc.)

Walter, C. (1969) Absolute Stability of Certain Types of Controlled Biological Systems. *Journal of Theoretical Biology, 23,* 39-52.

LECTURE V

In the preceding lectures, we have touched upon the basic problems of morphogenesis as they arise in biology, and some of the mathematical theories bearing on the generation of form and pattern in such systems. We have seen that these mathematical approaches primarily present us with *metaphors* for biological processes; the metaphors characterize large classes of systems which generate form and pattern in accord with some basic underlying dynamical principle, such as sorting-out, or the instability of a homogeneous state. Thus, each metaphor comprises a large class of different systems, any one of which may (or may not) be a model for a morphogenetic process of biological interest. But without further information, we have no way of extracting any specific system from the class which comprises the metaphor, in order to subject the system to closer analysis and empirical testing.

Although biological morphogenetic processes are clearly under the control of a genome, mathematical approaches to the generation of form are exclusively *epigenetic;* they do not require a genome for their operation. A reconciliation between genetic mechanisms and epigenetic morphological processes arises from the recognition that the genome serves to set initial conditions and constraints for higher-level dynamics. The crucial observation here is that the properties of interacting units which participate in these higher-level dynamical processes are, in general, *different* from those which are coded for by the genome. The question still remains: how are we to arrive at an integrated theory of form and pattern generation in biology, which will directly implicate the genome in the epigenetic mechanisms we have described? In the present lecture, we shall present some considerations bearing on this fundamental question.

In order to do this, we must retreat for a while to a consideration of first principles. In doing so, we will re-examine some of the material presented by other lectures, but from a rather different point of view. We shall consider basic epistemological questions which bear not so much on the properties of any particular mathematical formalism representing a biological (or any other kind of) process, but rather on the mechanisms required for the creation of such formalisms, and on the way in which such formalisms must be interpreted.

Our point of departure will be the fundamental property of genome-directed epigenetic dynamics, namely that the properties of the units entering into the dynamics are generally different from those coded for directly by the genome. If this assertion is accepted, it means, in particular, that all of the morphogenetic units must admit of at least *two different kinds* of descriptions: (a) a description in terms of the properties relevant to the coding, and (b) a description in terms of the properties relevant to the epigenetic dynamics. Furthermore, these descriptions must be different from each other.

Thus, the first fundamental epistemological question which we must consider is: what constitutes a description? Then we must ask how it is possible for the same system to admit different modes of description. And finally, what relations may exist between alternate modes of description of the same system, and how may they be integrated to obtain a fuller picture of the system?

What constitutes a system description, and how can such a description be obtained? Obviously, the empirical ingredients of any system description arise from *observation,* the interaction of the system with an observer or measurer. The theory of observation, or measurement, has played a central role in physical theory for many years, and never more so than since the development of quantum theory. Although the theory of microscopic measurement is still very much an object of controversy among physicists, nevertheless there are a number of basic principles which are universally accepted as applicable to all theories of measurement; these are:

1. any observations may ultimately be analyzed into one or more numerical-valued observations;

2. the only information which can pass between systems as a result of their interaction arises from such numerical-valued observations.

If we accept these postulates, provisionally, we obtain a unified way of deriving and representing system descriptions. We shall briefly review this procedure; fuller details are available elsewhere (Rosen, 1973).

If we regard any system as an abstract set of instantaneous states, then the first postulate above asserts that any observable property of a state may be represented by a set of numbers; these numbers in turn represent the outcomes of observing the state with instruments, which present their readings in numerical form. Thus, the action of each measuring instrument can be regarded as defining a mapping from the set of all the states of the system to the set of real numbers. We shall call such a mapping an *observable* of the system. The two postulates then assert that the only information which we can obtain about systems, and that systems can obtain about each other, ultimately devolves upon the evaluation of certain of the observables of the system on its states.

Conversely, at least in classical physics, it is held that any real-valued mapping defined on the set of states represents an observable i.e. corresponds to a measurable quantity pertaining to the states, in such a way that a measuring instrument could be constructed which would induce precisely that mapping. In microphysics, this kind of sweeping identification of observables with real-valued mappings apparently cannot hold; there exist apparent "super-selection rules" which limit the class of mappings which can represent observables. We are, nevertheless, entitled to say that there is a correspondence between the observables of a system, which carry the only information we can obtain regarding the system and its states, and a family of real-valued functions defined on the set of states.

Therefore, a description (or more accurately, a state description) of a system corresponds to specifying a family of observables and identifying the states of the system with the values of the mappings in this family. Thus, if $F = \{f_1, f_2, \ldots\}$ is a family of observables, we will identify σ, a state of the system, with the set of numbers $\{f_1(\sigma), f_2(\sigma), \ldots\}$. Typically, such a description is redundant, in the sense that some of the $f_i(\sigma)$ are determined as functions of others, and hence can be computed if the others are known. In ordinary dynamical systems theory, it is supposed that there is a finite set of observables of which all others are functions; these then comprise a family of *state variables* for the system, and we proceed to identify the states of the system with the set of numbers obtained by evaluating the state variables on the states.

It must be noted that this mode of system description is contingent upon a choice of F, the set of observables under consideration. If we chose a different set of observables, say $G = \{g_1, g_2, \ldots, g_n, \ldots\}$, we would generally obtain a different mode of state description. Indeed, there are as many modes of state description as there are subsets of the set of all observables of the system.

Conceptually, the choice of a set F of observables, around which a state description is built, amounts to the isolation of a *subsystem* of the original system. It also represents a process of *abstraction,* in the following sense: instead of employing the full set of observables which represents the system, we choose only a sub-family of these, and effectively throw away the rest. We are then left with an abstract subsystem which possesses only those inter-active capabilities which depend on the observables we have retained in our system description. Yet another way of describing this process is the following: we have made a *model* of the real system, by isolating a particular fraction of the interactive capabilities of the system, and discarding the rest. The model obtained by this process of abstraction will then behave like the real system precisely in those interactions which depend only on the retained observables; for all other interactions, the real system will behave differently from the model.

It should be noted that this process of abstraction, or model formation, or subsystem isolation (these terms are equivalent) is not restricted to a mathematical formalism, but permeates experimental biology as well. When a molecular biologist isolates a particular molecular fraction from a cell for analysis, that fraction represents an abstract or model cell, in precisely the sense we have described above. Each fraction he can extract will represent such an abstract cell, and each will have a different mode of state description, representing some fraction of the interactive capacities of the original intact cell. Thus, our discussion has been perfectly general and applies equally well to empirical or experimental, as well as to theoretical or mathematical, modes of analysis.

Let us now introduce another concept into our considerations. We can see immediately that there is a relation between the families of observables which comprise the modes of observation of a real system, and our view of the *complexity* of the system. More specifically, if we can only interact with the system in a single way, involving a single family F of observables, then we will regard that system as *simple*. Most of us, for instance, would regard a stone as a simple system, because we interact with it in ways which can be comprehended within a single mode of description. The more ways we have of interacting with a system, the more complex the system will seem to us. Therefore, to most of us, a cell is complex because we can interact with it in an enormous multiplicity of ways, each of which involves a different subsystem of observables. But to a geologist, who can typically interact with a stone in many ways not available to most of us, that stone which we consider simple can seem endlessly complex. Thus, we arrive at the following conclusions: (a) complexity, by itself, is not an inherent system property but is contingent on the number of ways available to us for interacting with the system; (b) each mode of interaction depends upon the isolation of a particular subsystem, characterized by a particular family of observables F of the system; (c) each family of observables gives rise to a different mode of description i.e. to a different model, of the system. Therefore, a complex system is one in which we have empirical access to many such subsystems; a simple system is one in which we have empirical access only to one or a few such subsystems.

Thus, complexity and the availability of alternate descriptions are two sides of the same coin. They are intimately related, on the one hand, to sets of measurements or observational processes available to us which serve to isolate or specify particular subsystems; on the other hand, they must play a crucial role in the genetic direction of epigenetic processes, where the interactive capabilities exploited in genetic coding are different from those exploited in epigenetic dynamics. Thus, we are led to consider the interrelationships between alternate descriptions, which are in effect descriptions of different subsystems of the same underlying system. There are many ramifications of this crucial problem, some of which have been discussed elsewhere (cf. Rosen, 1968, 1969); here we shall only consider a few of those which relate directly to morphogenesis.

Let us recall the basic situation. We have a class, Σ, of abstract states of a system and two different modes of description of these states, arising from two different families, F and G, of observables of the system. In mathematical terms, each of these families induces an equivalence relation on Σ in the following way: we shall call two states $\sigma_1, \sigma_2 \in \Sigma$ *equivalent* under a family of observables $F = \{f_1, f_2, \ldots\}$ if and only if $f_i(\sigma_1) = f_i(\sigma_2)$ for each $f_i \in F$. If we denote the equivalence relation so induced by R_F, then the representation of the system induced by F is the set of equivalence classes Σ/R_F; the representation itself involves identifying each $\sigma \in \Sigma$ with its equivalence class $[\sigma] \in \Sigma/R_F$, and thence with the set of values $\{f_i(\sigma)\}$, $f_i \in F$. It is the set of values which gives rise to "the state space" under this mode of description, as we have indicated above. Of course, we can do the same thing for the set of observables G, thereby giving rise to an alternate description of Σ through the set Σ/R_G.

Now any state space representation carries along with it a metric topology, by means of which we can say that two "states" are close.

Thus, in the F-description, we would say that two states $\sigma_1, \sigma_2 \in \Sigma$ are close if and only if $f_i(\sigma_1)$ and $f_i(\sigma_2)$ are close for each $f_i \in F$. This metric, of course, applies only to the set Σ/R_F, but by *abus de langage* we typically apply it to the set Σ itself; indeed, if we have only one mode of description for Σ, through the single family of observables F, we can do this in a completely consistent way. But when we have *two* descriptions, through the families F and G respectively, we may no longer be able to do this in a manner consistent with both descriptions. In particular, two "states" which are close under the F-description need not be close under the G-description. It is precisely this fact which gives rise to the concepts of genericity and bifurcation which have been applied by Thom (1972) in his general study of morphogenesis. Let us see, in more detail, how this comes about.

We have, in effect, a diagram of mappings of the following form:

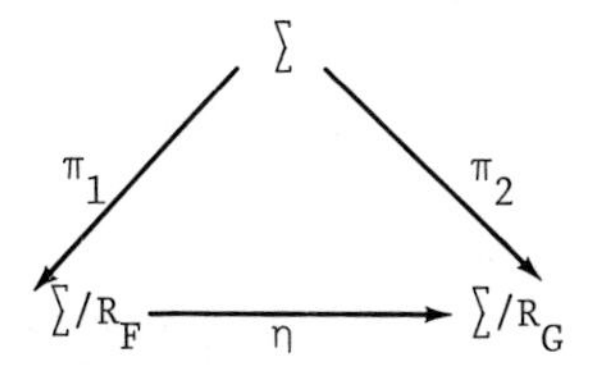

Here, the mappings π_1, π_2 are the natural projections of Σ onto the corresponding sets of equivalence classes. The mapping η is defined as follows: let $\sigma \in \Sigma$ be a representative chosen from the set $[\sigma]_F$ in Σ/R_F. Consider the corresponding equivalence class $[\sigma]_G$ in Σ/R_G, and write $\eta([\sigma]_F) = [\sigma]_G$. Clearly, the map η will, in general, depend upon the choice of representatives selected from the equivalence classes. But let us suppose that one such η has been chosen (corresponding, empirically, to the assumption that we can only look at one state $\sigma \in \sum$ at a time). Since both $\sum/R_F$ and $\sum/R_G$ possess metrics, we can ask for those classes at which the (purely set-theoretic) map η is continuous. These are then the *generic points* of the F-description with respect to the G-description; a generic "state" is then precisely one at which all "states" sufficiently close to the generic state in the F-description will map onto states which are close in the G-description. The complement of the generic states is the *bifurcation set;* at a bifurcation point, states which are close in the F-description will not, in general, be close in the G-description.

It should be noted that we can interchange F and G in the above diagram, and obtain, dually, the generic and bifurcation points of the G-description with respect to the F-description.

Thus, the concepts exploited by Thom arise quite generally when pairs of alternate descriptions of the same system are compared. For instance, in the classical definition of the structural stability of dynamical systems, the set $\sum$ can be taken as the set of all dynamical systems of the form

$$\frac{dx_i}{dt} = \alpha_i(x_1,\ldots,x_n), \quad i = 1,\ldots,n$$

defined on a particular manifold co-ordinatised by the state variables $x_1,\ldots,x_n$. The F-description can be chosen in terms of the properties of the functions α_i which specify the system dynamics; two systems will be "close" if their corresponding functions α_i are close in some appropriate topology. In this case, we can take the sets $\sum$ and $\sum/R_F$ as identical;

thus, the map η is always well-defined. In the G-description, however, we shall consider the properties of corresponding trajectories of the system in terms of their separation as measured in the topology on the manifold of states. Two dynamical systems in this description are "close" if their corresponding trajectories are always "close" in this topology. A dynamical system is called structurally stable if and only if any other dynamical system, sufficiently close to it in the F-description, is always "close" to it in the G-description. Thus, the classical ideas of structural stability arise as a special case of a far more general situation in which two alternate descriptions of the same system are compared with each other in a particular way.

Intuitively, the generic points are points at which the two descriptions essentially agree; the bifurcation points are those at which the two descriptions disagree. Thus, bifurcation reveals the incompleteness of a particular mode of state description, and it is at the bifurcation points that the alternate description supplies new information with respect to the system. The relationship between bifurcation and information has been discussed elsewhere (Rosen, 1975a).

There are a number of other important consequences arising from these notions of alternate descriptions which should be mentioned here. One of these has to do with *emergence,* the name given to the appearance of entirely new structures and behaviors during morphogenesis. As we saw in Lecture III, emergence is characteristic of epigenetic processes; emergent novelty is also one of the main characteristics of evolutionary processes. Phenomena of emergence have always seemed mysterious because they indicate situations in which one mode of system description must be replaced by another; typically no explanation within a single mode of description can be found (cf. Platt, 1961).

More precisely, since every mode of system description is an abstraction from which interactive modes present in the system have been subtracted, any description is useful only in dealing with interactions and their effects, which pertain to the features of the system retained in the abstraction. Emergent phenomena, when seen from the perspective of such a description, pertain to interactions not comprehended within the description but which ultimately affect the variables in that description. Thus, in principle, emergent phenomena are not predictable within a single system description, since the basis for them has been abstracted away in generating the description. All that can be done, within a single mode of description, is to analyze the sensitivity of the behavior of the model to perturbations of its dynamics; this kind of study has been initiated by Thom. However, the source and character of such perturbations, which is intrinsic to a real developing system, must be extrinsic to any single mode of description of that system.

A related feature of multiple descriptions of complex systems is the concept of *error*. Intuitively, we feel that simple systems, such as those arising in mechanics, cannot make errors; indeed, von Neumann (1956) and other have argued that the concept of error requires a threshold of complexity in order to be meaningful. In terms of the viewpoint we have been developing, an *error* is recognized in terms of a deviation between the actual behavior of a real system and the behavior predicted by a simple model of the system. In greater detail, the actual behavior of a real system represents the sum total of its interactions with the environment and its intrinsic behavior. In any simple model, or abstraction, we have seen that most of the interactive capacities of the real system have been abstracted away, and hence inevitably lost. Thus, any interaction depending on such missing capabilities will, in general, cause a deviation between the behavior of the real system and the behavior of the model; this deviation provides a basis for the concept of error. We often attempt to replace the missing interactive capabilities by superimposing a stochastic variation on those retained in our description, but from the present point of view this is a highly imperfect precedure. Thus, we see that the concept of error, like that of emergence, arises from attempting to replace the total interactive capacity of a real system by that possessed by a subsystem. Such a viewpoint throws a different light on the problem of "noise" or "error" in biological systems, and in physical systems generally. For greater details, see Rosen (1975b), where the relation of these ideas to physiological processes and to evolution are developed in greater detail than is possible here.

Following this prolonged excursion into some of the basic epistemic aspects of system modelling, let us return to the fundamental question of the genome and its relation to the epigenetic generation of pattern and form. Once again, we repeat that the very concept of genetic coding requires us to utilize at least two alternate descriptions of the morphogenetic elements; one description pertains to the interactions, bearing on the coding itself, and another pertains to the epigenetic dynamics not directly coded for. The basic ingredient in all of these descriptions is the choice of suitable subfamilies of observables pertinent to these interactions and hence the choice of appropriate subsystems or models. The observation process itself is crucial as the underlying ingredient of the descriptions we require. Therefore, let us turn to the properties of a formalism incorporating these various aspects and try to obtain some insight into what a genome-directed morphogenetic theory might be like.

It seems reasonable to suppose, at the outset, that such a theory will have to be one in which patterns of genomic activity are translated, or read out, into corresponding patterns in the populations of cells controlled by the genome. That is, we will need to find a kind of transduction mechanism between patterns of genome activity and the corresponding patterns which they generate, through epigenetic processes, at the phenotypic level. Each epigenetic process, in turn, must be characterized by an appropriate set of observables which corresponds to the relevant *features* displayed by a pattern of activity, and is coupled by the transducing mechanism.

This problem can be recast, usefully, in another language. On the one hand the problem requires the *recognition* of a pattern of genomic activity, while on the other hand this recognition is reflected in the *generation* of a corresponding pattern at the phenotypic level. It is possible to argue that one cannot, in fact, recognize a pattern in a given system without simultaneously generating a pattern in some associated system. Conversely, one can argue that the generation of a pattern in a system is the result of the recognition of a pattern through interaction with an associated system. If this is true, then pattern recognition and pattern generation become inseparable, and this strongly suggests that the problem of genome-directed morphogenesis in biology can be considered in this way.

Notice, also, that the problem of pattern recognition is, in a certain sense, a generalization of the problem of measurement in physics. Whereas measurement, as we have seen, restricts attention to those interactions resulting in numerical values, and classifies systems (and the states of systems) on this basis, pattern-recognition typically involves classifications based on "features" of an apparently more qualitative type. In abstract terms, this is the only difference between the notions of measurement and of pattern recognition. Moreover, if it is true that every "feature" can itself be characterized in terms of numerical-valued functions, or observables, then the two theories are conceptually identical. The explicit confluence of the two approaches can be found at the molecular level, in problems of conformation and specificity; for instance, the specificity of enzyme-substrate interactions, or antigen-antibody interactions, arises from features generated by the molecular architecture of the substances involved. These features fall within the province of physics and hence, by the fundamental postulates concerning measurement, should be characterized in terms of numerical-valued observables defined on the sets of states of the molecules involved. It must be explicitly recognized, however, that the usual physical description of these molecules typically utilizes only a small fraction of the available observables, and therefore may fall outside the above descriptions (cf. Rosen, 1960; Comorosan, 1975).

The relation between measurement and pattern recognition suggests that there is a common formal framework in terms of which both processes can be represented; such a formal framework would then be a general theory for all types of classification or recognition processes. Let us consider how we might go about constructing such a framework.

All cases of measurement, recognition and classification involve at least two basic ingredients: (a) a set S of elements which we wish to classify, discriminate, or measure according to a set of features exhibited by the elements of S; (b) a system M, variously called an observer, measuring device, or classifier. The system M can be coupled to the elements of S; as a result of this coupling or interaction, the system M, on the basis of its own autonomous dynamics, will move along a trajectory which has asymptotic properties that are related to the elements of S with which the system M was coupled.

In order for a system M to function as a classifier, or measurer, or pattern recognizer, its own autonomous dynamics must be subject to the following constraints:

1. The system M must posses a (conditionally stable) *zero-state* σ_0, such that, in the absence of interaction between M and the elements of S, the system M will remain in σ_0 indefinitely.

2. An interaction between $\sigma_0 \in M$ and a state $s \in S$ will displace, or perturb, M to a new state. In this state, the autonomous dynamics of M will carry M to a well-defined asymptotic behavior.

3. Two states sufficiently close in the topology of the state space of M must be carried to the same asymptotic behavior. Thus, each asymptotic behavior of M will correspond to an open subset of the state space, and the (closure of the) union of all these open subsets must be the entire state space. In other words, if $U_1, U_2, \ldots$ are the open subsets in question, then any neighborhood of σ_0 intersects each of the sets U_i.

Intuitively, then, a dynamical system satisfying these properties can serve as a measuring device, or classifier, or recognizer, of patterns in the elements of the set S to which it can be coupled. According to the above properties, the system M can classify *its own states*, on the basis of its own autonomous dynamics, into a discrete set of asymptotic behaviors; we shall call such systems *autonomous state classifiers*. The trajectories of the *binary classifier*, which classifies its states into two asymptotic behaviors, and the *ternary classifier*, which classifies its states into three such behaviors, are shown below:

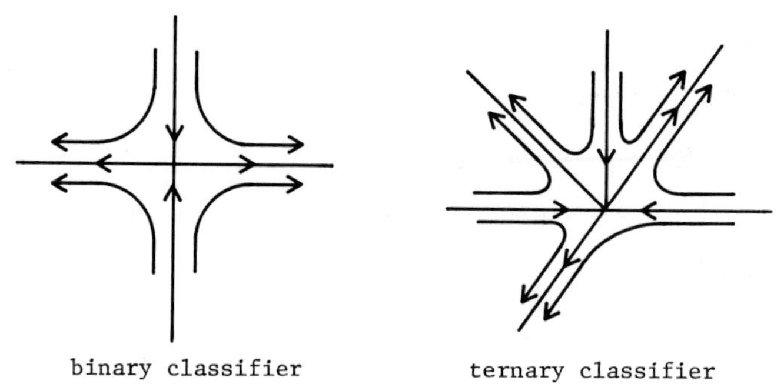

It can be shown (Rosen, 1972a, 1972b; Engel, 1973) that every N-ary classifier, for any N, must have trajectories of this type. For $N = 2$, the classifier is essentially a saddle point; for $N > 2$, the trajectories are like saddle-points. It should be noted that these systems are not only non-linear, but cannot be linearized in any neighborhood of σ_0.

Intuitively, interaction with a pattern $s \in S$ will displace the classifier to some new state lying in one of the open regions U_i. Which of these regions will be entered will depend entirely on s, and on the mode of coupling between S and M. Once a region has been entered, the autonomous dynamics of M will drive the system to a unique asymptotic behavior. This asymptotic behavior can be regarded as the pattern *generated* in M as a result of the recognition of a particular feature of an element

$s \in S$.

If we regard the elements of S as representing patterns of genetic activity, and the behavior of M as representing corresponding phenotypic patterns, we see that we have most of the ingredients required for a genome-directed theory of morphogenesis. The genome in this case sets initial conditions for the autonomous classifier dynamics, since a genomic pattern determines a unique displacement of the classifier from its zero-state to one of the regions U_i dominated by a particular attractor. Moreover, the interactions between the genome patterns $s \in S$ and the classifier are different from those involved in the classification dynamics. In this scheme, the classifier dynamics constitute the epigenetic component while interactions between the classifier and the genomic patterns represent the genetic control.

We can expand this basic picture in a variety of ways. For one thing, we have supposed that the elements of S are static objects, possessing no intrinsic dynamics, and that the interactions of the elements of S with M are of an *impulsive* character. These assumptions can be considerably weakened. For instance, the interaction between S and M will not only displace M away from its initial state, but will, in general, also displace S. This displacement can be regarded as *an observation of the state of the classifier by the genomic pattern*. Thus, the state of the genome itself will be modified by the interaction with M, and if we assume an intrinsic dynamics on S, such as that implicit in the kinds of genetic networks described in Lecture III, we will inevitably find that such changes of genetic pattern can be interpreted as the "turning on" of genes previously repressed and the "turning off" of genes previously active.

Moreover, as far as the classifier M itself is concerned, it turns out (cf. Rosen, 1972b) to be generally possible to divide its activity into two parts, one concerned with the underlying classifier dynamics, and the other concerned with the setting of *rate constants* for this dynamics. In greater detail, let us suppose we have an arbitrary dynamical system; for the sake of convenience we represent it as the two-dimensional system

$$\frac{dx}{dt} = \theta_1(x,y),$$

$$\frac{dy}{dt} = \theta_2(x,y).$$

The functions θ_1, θ_2 which define the dynamics are real-valued functions on the set of states, and hence are observables of the system. The set of all observables is a linear space, and so θ_1, θ_2 can be expanded in any convenient complete set of basis functions $\Psi_i(x,y)$, say. Thus, we can write the dynamics in the form

$$\frac{dx}{dt} = \sum_{j=1}^{\infty} \alpha_{1j} \Psi_j(x,y)$$

$$\frac{dy}{dt} = \sum_{j=1}^{\infty} \alpha_{2j} \Psi_j(x,y)$$

where α_{1j} and α_{2j} can be interpreted as specific *rate constants*. If we now couple such a system to a classifier M, so that these rate constants become explicit functions of the classifier state, we can use the classifier to derive many other systems, and thus add higher levels of epigenetic control. Such a picture is in accord with the idea that the primary effect of the genome is the manipulation of reaction rates.

This kind of picture is consistent with all the requirements for a genome-directed theory of pattern generation, and all of the specific morphogenetic mechanisms mentioned in the earlier lectures can be incorporated into such a picture in a natural and satisfying way. However, there are several inherent drawbacks. We shall discuss some of these from a general, conceptual point of view; by so doing, we may make contact with further aspects of the relation between mathematical formalisms and the situations they represent, and at the same time suggest some important directions for further work.

The first general objection which can be levelled against the scheme we have presented is that it remains a metaphor and not a model. Thus, in common with all metaphors, it embodies a large class of diverse morphogenetic systems related only by a common principle of organization, and all of the considerations relating to dynamical metaphors which were presented in the preceding lecture apply here as well. In particular, there is the universal difficulty of finding explicit experimental correlations with specific morphogenetic processes. However, as we have argued above, we should not expect traditional ideas of verification and falsification, which apply to models, to be meaningful at the metaphorical level. The usefulness of a dynamical metaphor as a basis for scientific explanation is found at a different level, and a metaphor must be considered on its own terms.

Another difficulty, which is also closely connected with the problem of relating metaphors and models to experiment, should also be discussed. This difficulty concerns the relation between the variables appearing in formal dynamical treatments of morphogenetic processes, and those observable quantities of real morphogenetic systems which we find it convenient to measure. In conventional model-building, where experimental data is taken as the starting-point for the model, such observables are incorporated directly into the mathematical formalism, and no problem arises. But at the metaphorical level, or in the attempt to analyze gross morphogenetic processes in terms of underlying dynamical ones, there is no guarantee that the quantities appearing in the analysis will have any immediate interpretation in terms of quantities which are easy to measure. This question of interpretability is a crucial one, and we will conclude these lectures by briefly discussing several aspects of it in the light of what we have already said.

Let us recall that the purpose of analysis is, after all, to decompose a system too complex to be dealt with directly into a family of simpler subsystems. We then seek to understand the properties of the original system of interest in terms of the properties of the subsystems. As we saw in Lecture I, the cell theory of organisms, or more generally the decomposition of any morphogenetic process into the behaviors of morphogenetic subunits, represents an instance of this kind of analysis. The subsystems considered in any of these cases, however, have a very special property: they are physical units, which may be physically separated from the total system, and physically studied in isolation. Thus, there is clear empirical motivation for regarding information about these subsystems as pertaining directly to the total system from which they were isolated.

But there exist many kinds of subsystems at least as appropriate for the analysis of complex systems activity which cannot be *physically* isolated from the total system. A subsystem such as a pool, or compartment (which represents a reservoir of chemical reactivity), or an operon, or an active site in a catalytic protein, is an example of a functionally meaningful subsystem which we cannot physically separate or isolate from the total system. Yet these subsystems represent meaningful units of interactive capacities which are indispensable for an understanding of crucial biological behaviors. Although such subsystems are not embodied in a physical unit, or particle, which can be isolated by physical means, this does not make them less real or significant for the purposes of analysis.

As we noted above, the purpose of analytic technique is to enable us to understand the interactive capabilities of a complex system in terms of the interactive capabilities of subsystems. We have seen that such interactions can only occur, according to the hypotheses ordinarily made in science, through the values assumed by particular observables on the states of the system. To study any such interaction, then, we must decompose the system into subsystems which will allow the observables responsible for the interaction to be reconstructed from the observables of the subsystems. It is clear that, in general, an arbitrary decomposition of the original system into subsystems need not have such a property. Any decomposition which lacks this property will be useless for studying the interactions in question, for it destroys the basis on which the interactions depend. In particular, there is no guarantee that a decomposition based on a particular set of empirical techniques, which characterize subsystems in terms of physical separability, will be useful in the study of a particular mode of biological interaction, however "natural" the decomposition may otherwise seem.

Let us look at a specific example. Consider a system of chemical species $x_1,\ldots,x_n$ interacting with each other according to first-order kinetics. Then according to the Law of Mass Action, the system will be governed by a set of dynamical equations of the form

$$\frac{dx_i}{dt} = \sum_{j=1}^{n} a_{ji}x_j, \quad i = 1,\ldots,n.$$

The state space of the system is the first orthant of real Euclidean n-space, and the variables of interest, those which we can conveniently measure empirically, are the concentrations x_i. Each x_i specifies a subsystem of the total system, and one which can be physically separated from the total system.

But we cannot understand this system by isolating the subsystems x_i and studying them in isolation. Rather, we must proceed formally, by introducing a new coordinate system into E_n which is more naturally related to the dynamics governing the system. In particular, we define *new observables*

$$u_i = \sum_{j=1}^{n} \beta_{ij}x_j, \quad i = 1,\ldots,n,$$

linearly related to the original state variables, where the coefficients β_{ij} are so chosen that the system matrix assumes a particularly simple canonical form. If we can diagonalize the system matrix in this fashion, the system equations become

$$\frac{du_i}{dt} = \lambda_i u_i, \quad i = 1,\ldots,n,$$

which can be solved immediately; through an inverse transformation to the original variables x_i, we can find out what we want to know about the original system.

The conceptual significance of this procedure is the following. The variables u_i serve to *decouple the system*, so that instead of dealing with a highly interactive system, in which the rate of change of each variable depends on all of the others, we have a situation in which the rate of change of each variable depends only on that variable. If we wanted to understand our original system by decomposing it into subsystems, the subsystems specified by the u_i would be the ones to choose. This is true even though we cannot measure the u_i directly, nor physically separate the system in these terms.

Of course, the u_i are regarded as convenient mathematical fictions, even though they are, by definition, system observables (i.e. numerical functions of the state variables) and therefore could in principle be measured directly. The point we wish to emphasize here is that the subsystem specified by the u_i is the natural basis for analyzing the given system, even though it is the x_i which we can measure and physically isolate. Stated another way, decompositions which look "natural" to us from the standpoint of easy measurability may be most unnatural from the standpoint of understanding global modes of interaction, and vice-versa.

It will be observed that we used a procedure of exactly this kind in analyzing the Rashevsky-Turing metaphor in Lecture II. The example we considered was couched in the language of "cells", and chemical species, or morphogenes, flowing between them. But in order to solve the system, we had to introduce new state variables which decomposed the system of "cells" into an entirely different family of subsystems, in which the "cells" were not visible. The necessity of doing this shows that cellular organization does not provide the natural analytic framework for understanding the dynamical interactions which it generates. Furthermore, an exclusive preoccupation with the cellular organization would actually make it impossible to understand the morphogenetic capabilities of the system as a whole. We must pass to a different description of the system, one which *commutes* with the dynamics (in the sense that its subsystems decouple that dynamics), in order to actually understand what is going on. The conditions under which this procedure can be done represent, we feel, an important area of pure mathematical research (cf. Rosen, 1972c).

These elementary examples indicate, in yet another way, the necessity for multiple descriptions of the same system which we have emphasized at great length in this lecture. But most important for present purposes, these considerations show that we must be ready to take a more coordinate-free viewpoint with respect to our morphogenetic models. Of course, we must be ultimately interested in results couched in terms of quantities which it is possible, and convenient, for us to measure. But we must remember that these quantities comprise only a small fraction of the available interactive capabilities of the systems with which we deal, and often they are not appropriate for the analysis of such capabilities. *What is important in these cases is not so much how we can see these systems, but how they can see each other.*

In metaphoric terms, these considerations mean that in the process of abstracting individual models from a given dynamical metaphor, we must be prepared to find that these models will be couched in terms not immediately comparable to experiment. For example, we must be prepared to recognize cellular organizations in dynamical models which do not, at first sight, seem to be organized in terms of cells at all. We must learn to be at home in situations describable both in terms of subsystems which are dynamically natural and in terms of subsystems which are experimentally natural; unfortunately, the two seldom seem to coincide. We are just beginning to appreciate the fact that many of the difficulties of morphogenesis, and of biology in general, spring precisely from this lack of coincidence.

REFERENCES

Comorosan, S. (1975) Biological Observables. In *Progress in Theoretical Biology, Volume 4*. (New York: Academic Press, Inc.)

Engel, A. (1973) Plane Autonomous State Classifiers. Ph.D. Dissertation. Buffalo: State University of New York.

von Neumann, J. (1956) Probabilistic Logics and the Synthesis of Reliable Organisms from Unreliable Components. In *Automata Studies*, C. Shannon and J. McCarthy, eds. (Princeton: Princeton University Press)

Platt, J.R. (1961) Properties of Large Molecules which go Beyond the Properties of their Chemical Subgroups. *Journal of Theoretical Biology, 1*, 342-358.

Rosen, R. (1960) A Quantum-Theoretic Approach to Genetic Problems. *Bulletin of Mathematical Biophysics, 22*, 227-255.

Rosen, R. (1968) On Analogous Systems. *Bulletin of Mathematical Biophysics, 30*, 481-492.

Rosen, R. (1969) Hierarchical Systems in Biology. In *Hierarchical Structures*. L.L. Whyte, A.G. Wilson and D. Wilson, eds. (New York: American Elsevier Publishing Co.)

Rosen, R. (1972a) Autonomous State Classification by Dynamical Systems. *Mathematical Biosciences, 14*, 151-167.

Rosen, R. (1972b) Further Comments on Autonomous State Classifiers and an Application to Genetics. *Mathematical Biosciences, 14*, 305-310.

Rosen, R. (1972c) On the Decomposition of Dynamical Systems into Non-Interacting Subsystems. *Bulletin of Mathematical Biophysics, 34*, 337-341.

Rosen, R. (1973) Some Comments on Observability and Related Topics. *International Journal of Systems Science, 4*, 65-76.

Rosen, R. (1975a) Structural Stability, Alternate Descriptions and Information. Submitted to *Journal of Theoretical Biology*.

Rosen, R. (1975b) Complexity as a System Property. To appear in a forthcoming volume on system complexity, edited by J. Cornaccio.

Thom, R. (1975) *Structural Stability and Morphogenesis*. (D.H. Fowler, trans.) Reading: W.A. Benjamin, Inc.

Catastrophe Theory and the Modelling of Biological Systems

Professor A.E.R. Woodcock

Department of Biology
Williams College
Williamstown, Massachusetts

CATASTROPHE THEORY - A GEOMETRICAL INTRODUCTION

Catastrophe Theory, proposed by Thom (1969), describes the behavior of a dynamical system in terms of the maxima and minima of the associated potential energy function. The minima represent stationary or (quasi-) equilibrium conditions for the energy function and serve as states of attraction for the dynamical system, while the maxima act as repellor states. The potential energy function V is parameterized by a manifold C, the *control space*, on a manifold X, the *behavior space*. The set

$$M = \{(c,x) \in C \times X \mid \nabla_x V(c,x) = 0\}$$

defines the *catastrophe manifold* when V is differentiable everywhere (Woodcock and Poston, 1974a). A *catastrophe* is a singularity of the map

$$\chi : M \to C$$
$$(c,x) \to c$$

The catastrophe manifold is, therefore, a hypersurface generated as the loci of the maximum and minimum values of the associated potential energy function; a point on the surface represents a stationary state of the system. The transition of the system from initial to final states may be represented as a trajectory of a point on the catastrophe manifold. Since the trajectory of a state point is determined by changes in the coordinate positions on the control space, these coordinates become the *control parameters* of the system. The values of these parameters for which transitions actually occur may be determined with reference to the restrictions imposed by certain *ad hoc* transition conventions (see for example, Thom, 1975;

Goodwin, 1971).

Thom (1969, 1972, 1975) has shown that for a maximum of four controlling parameters there are only seven possible families of transitions from initial to final states for the system. These seven are called the 'Elementary Catastrophes' since transitions between states result, in general, in the catastrophic annihilation of the stability of the initial state. In four of these seven, the so-called cuspoids, the potential energy well is one-dimensional while in the remaining three it is two-dimensional (Table I).

The potential energy functions may be expressed as polynomials of the form

$$V_{cuspoids} = \frac{x^n}{n} + \frac{ax^{n-2}}{n-2} + \ldots + rx \quad (1)$$

$$V_{umbilics} = fn(x,y,t,w,u,v) \quad (2)$$

where $a,\ldots,r$ and t,u,v,w are the *control parameters* and x,y are the *behavior parameters* of the system.

TABLE I

The Elementary Catastrophes (after Thom [1969])

Catastrophe	Germ	Unfolding
Cuspoids		
Fold	$x^3/3$	ax
Simple Cusp	$x^4/4$	$\frac{ax^2}{2} + bx$
Swallowtail	$x^5/5$	$\frac{ax^3}{3} + \frac{bx^2}{2} + cx$
Butterfly	$x^6/6$	$\frac{ax^4}{4} + \frac{bx^3}{3} + \frac{cx^2}{2} + dx$
Umbilics		
Hyperbolic	$x^3 + y^3$	$wxy + ux + vy$
Elliptic	$x^3 - 3xy^2$	$w(x^2 + y^2) + ux + vy$
Parabolic	$y^4 + x^2y$	$ty^2 + wx^2 + ux + vy$

The Simple Cusp Catastrophe

The cusp catastrophe is associated with the following potential energy function:

$$V = \frac{x^4}{4} + a\,\frac{x^2}{2} + bx\ . \tag{3}$$

In this equation, a and b define the control space and x specifies the behavior space. The catastrophe manifold is a hypersurface in three-dimensional (x,a,b) space. The stationary values of the potential energy function are obtained by solving

$$\frac{dV}{dx} = x^3 + ax + b = 0 \tag{4}$$

for x (see Figure 1).

Equation (4) has at most three real roots (for negative a) and at least one real root under all conditions. For example, when $b = 0$, (4) has solutions $x = 0$ and $\pm\sqrt{-a}$. In the simple cusp these solutions represent two minima and one maximum of the energy function. Solving equation (4) for all pairs (a,b) traces out a surface in (x,a,b) space, the catastrophe manifold, generated by the loci of the maximum and minimum values of the potential energy function. Changes in the magnitudes of the control parameters may cause the system to assume different stationary states. The values of the controlling parameters a and b for which one or other of the stationary states disappears may be obtained from (3) by double differentiation and substitution for x.

The equation

$$4a^3 + 27b^2 = 0 \tag{5}$$

describes the *bifurcation set* of values (a,b) at which two of the

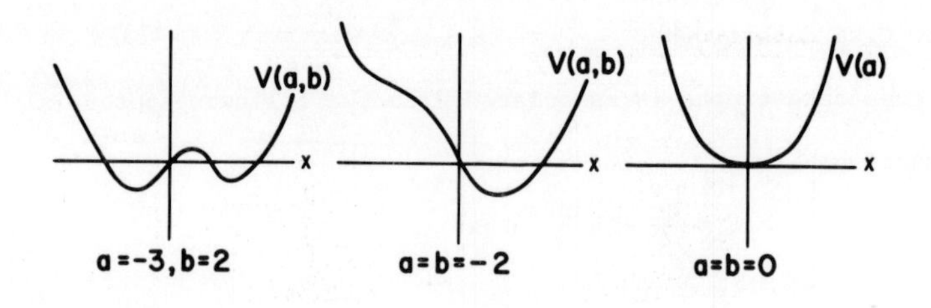

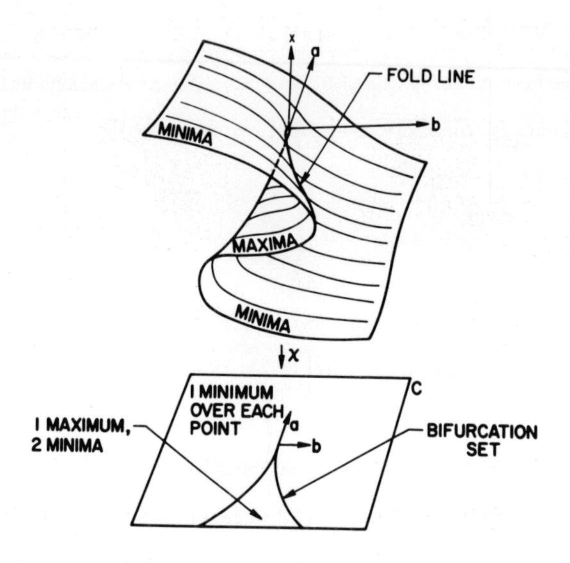

Figure 1 The potential energy well, associated manifold and bifurcation set of the simple cusp catastrophe. The potential energy well of the cusp catastrophe has, at most, two maxima and one minimum. The dual cusp exhibits two maxima and one minimum. The catastrophe manifold consists of a sheet of maxima separating two separate sheets of minima. These sheets are joined along the fold lines. The precise values of the parameters a and b for which fold lines occur are the bifurcation set.

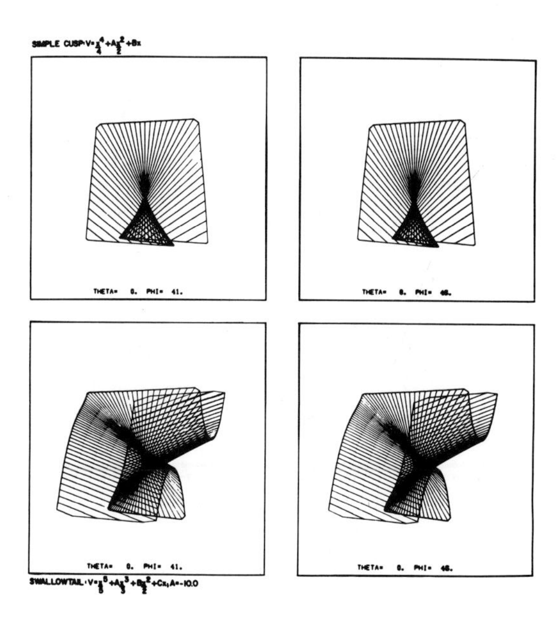

Figure 2 Stereographic pair pictures of the simple cusp and the swallowtail catastrophes. The three-dimensional nature of the catastrophe manifolds may be reconstructed by stereographic viewing of these pictures (cf. Woodcock, 1974a, 1974b).

stationary states vanish as two of the roots of equation (4) become complex.

The geometrical shape of the catastrophe manifold and the associated bifurcation set of the simple cusp and the other elementary cuspoid catastrophes has been explored using computer graphics techniques. Ruled line projections of the catastrophe manifold (Woodcock and Poston, 1974a) and stereographic pictures (see Figure 2, for example) (Woodcock, 1974a, b,c,d,e) have produced precise images of these surfaces and of their projections onto the control parameter space.

The Umbilic Catastrophes

The hyperbolic, elliptic and parabolic umbilics (see Table I) all have associated potential energy wells with two behavior dimensions; as a consequence, lines ruled on the catastrophe manifold surface are curved. The nature of these surfaces has been explored using computer graphics techniques (Woodcock and Poston, 1974b,c). Subsequent study of these surfaces using a more sophisticated technique employing a hidden-line method has produced less visually-ambiguous figures (Woodcock and Appel, to appear). These figures clearly reveal that the hyperbolic umbilic consists of a 'folded, sheet-like' surface; the elliptic umbilic is a hypercycloid with three cusps and a line of points of apparent self-intersection, while the parabolic umbilic defines an s-shaped surface with a line of apparent triple intersection points (see Figures 3,4, and 5).

Catastrophes of Higher Order

The geometry of the catastrophe manifolds of some non-elementary

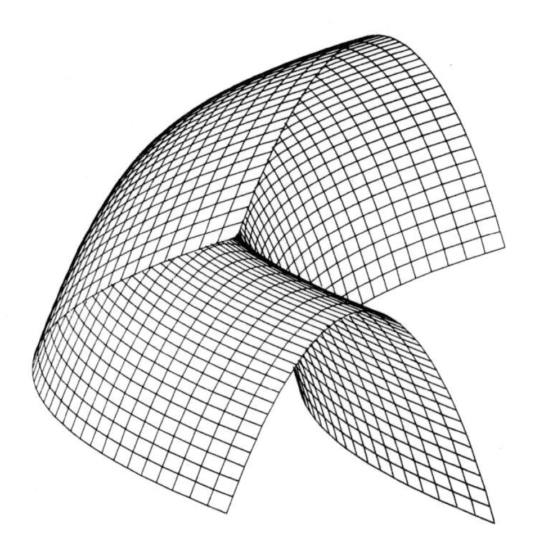

Figure 3 Hidden-line elimination picture of the hyperbolic umbilic catastrophe manifold (cf. Woodcock and Appel).

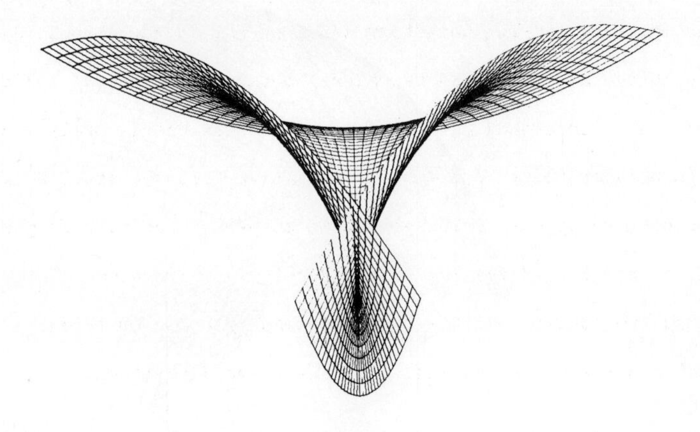

Figure 4 Hidden-line elimination picture of the elliptic umbilic catastrophe manifold (cf. Woodcock and Appel).

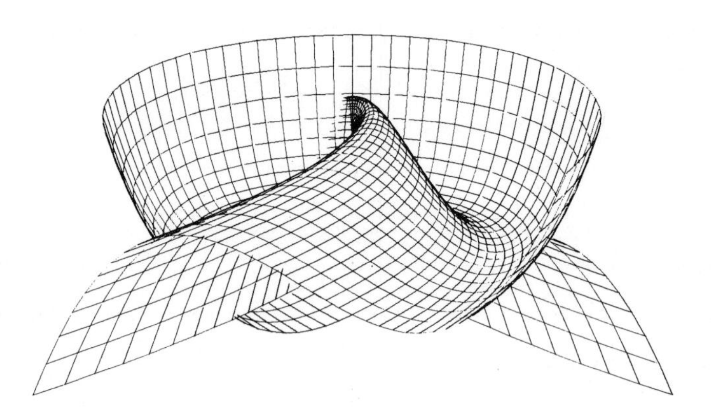

Figure 5 Hidden-line elimination picture of the compactified version of the parabolic umbilic. This technique of drawing removes the visual ambiguity that has occurred in previous presentations of this surface (cf. Woodcock and Appel).

catastrophes has been examined. Projections of the catastrophe manifold onto the control parameter space of the Reduced Double Cusp exhibits three- and four-fold symmetry (Woodcock and Poston, 1974d). The projected catastrophe manifold of the Reduced Double Swallowtail (see Figure 6) exhibits three-, four- and five-fold symmetry (Stewart and Woodcock, to appear). The geometric shapes of the catastrophe manifold of the Full Unfolding of the Double Cusp (Woodcock, Stewart and Poston, to appear) and the Anonymous and Symbolic Umbilics (Woodcock and Poston, to appear) have also been determined.

(1-3) and (3-1) Stable Catastrophe Manifolds

Recently Wasserman (1975, to appear) described a number of catastrophes with associated potential energy functions containing a more complicated parameterization of spatial and temporal control components. The bifurcation set associated with the function

$$V = x^4 + ux^2 + tx^2 + vx + t^2x \tag{6}$$

and

$$V = x^4 + ux^2 + t^2x^2 + vx + t^2x + wtx \tag{7}$$

occupies different locations on the (u,v) and (u,v,w) control parameter subspace, respectively, for different values of the parameter t (Woodcock, to appear).

Catastrophe Machines

Zeeman (1972) has devised a machine using elastic energy to demonstrate catastrophic behavior. In the machine linear changes in the parameters controlling the system can lead to discontinuities in its behavior. This machine has been analyzed in some detail (Poston

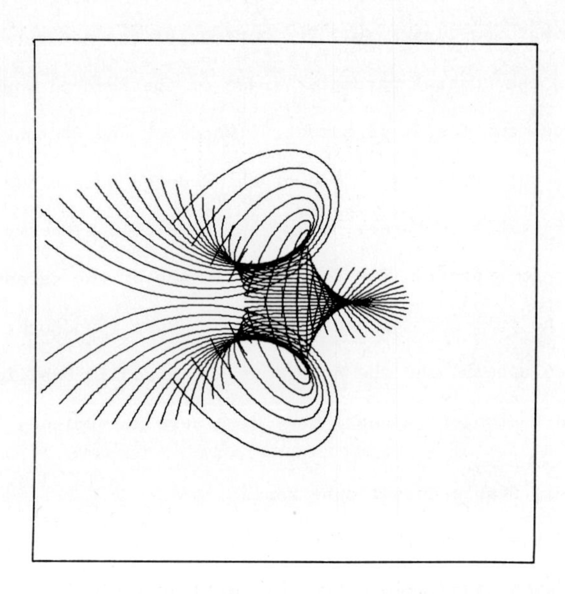

Figure 6 The reduced double swallowtail. The reduced double swallowtail also exhibits both three-fold and four-fold symmetry for other values of the control parameters (cf. Stewart and Woodcock).

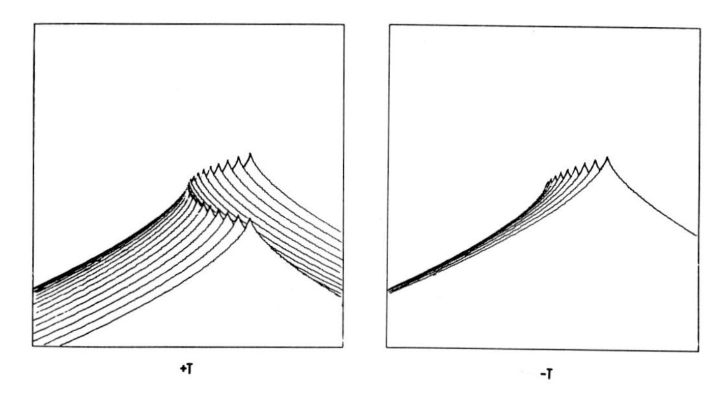

Figure 7(a) $V = x^4 + ux^2 + tx^2 + vx + t^2x$

Hidden line elimination pictures of the bifurcation set for 1-3 stable catastrophes (after Wasserman, 1975). The bifurcation set has been drawn for a fixed value of t and w and corresponds to a range of x for each particular line. The figures are drawn as if looking from the plus t direction towards zero and the minus t direction towards zero and enable the full magnitude of the relative motion of the bifurcation set to be appreciated (cf. Woodcock).

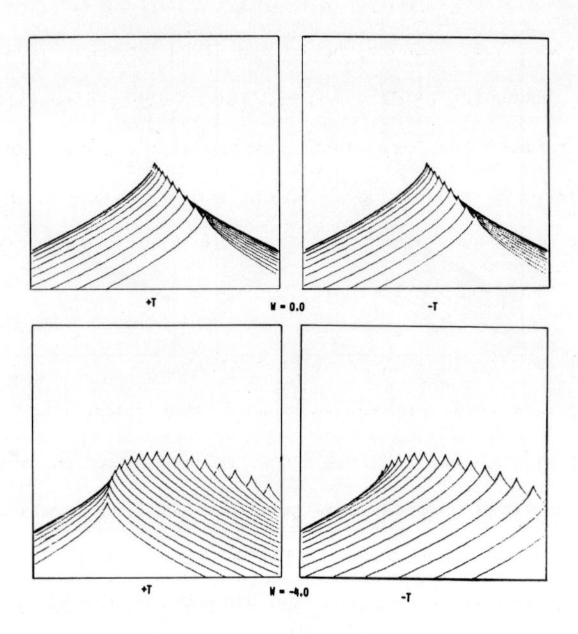

Figure 7(b) $V = x^4 + ux^2 + t^2x^2 + vx + t^2x + wtx$

Hidden line elimination pictures of the bifurcation set for 1-3 stable catastrophes (after Wasserman, 1975). The bifurcation set has been drawn for a fixed value of t and w and corresponds to a range of x for each particular line. The figures are drawn as if looking from the plus t direction towards zero and the minus t direction towards zero and enable the full magnitude of the relative motion of the bifurcation set to be appreciated (cf. Woodcock).

and Woodcock, 1973; Dubois and Dufour, 1974; Poston, 1975, to appear). Figure 8 shows the bifurcation set for a higher-dimensional catastrophe machine in which the long elastic string in the Zeeman MKI machine has been replaced by two strings. As the angle between these strings increases, the lower simple cusp develops into a butterfly-like envelope (Woodcock and Poston 1976).

Biological Implications of Catastrophe Theory

In formulating Catastrophe Theory, Thom (1969, 1972, 1975) was stimulated by the possible role which it could play in providing qualitative models of biological processes in general and of morphogenesis in particular. He pointed out that biological systems are very complicated and the task of providing exact information about all the possible variables of the system would be an extensive, if not impossible, undertaking. Under these circumstances, he suggested that a more useful approach would be to construct qualitative rather than quantitative models. These models would predict the types of changes that could occur rather than giving precise information concerning times when these changes would take place, for instance. For a restricted class of systems (such as those having an associated everywhere differentiable, potential energy function and also having at most four control parameters) elementary catastrophe theory may provide a suitable group of models for these purposes. Applications of catastrophe theory to problems in developmental biology have since been attempted; for examples, see Woodcock (1974 f) and Zeeman (1974, 1976).

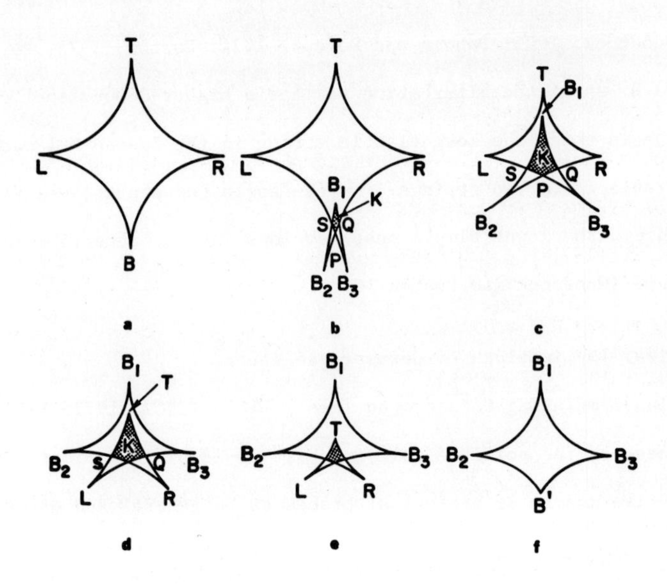

Figure 8 The bifurcation envelope for a higher dimensional version of the Zeeman Catastrophe Machine in which the lower elastic for the machine is a double strand instead of the single strand in the Mk. 1 machine. The ends of this strand are attached at two symmetric points on either side of the vertical midline of the picture. As the separation of the two end points of the rubber strand is increased the envelope changes from one of four cusps; first the lower cusp (labelled B in the picture) splits, then it forms a butterfly point and finally it becomes three cusps, B1, B2, and B3. T and B are normal cusps and L and R are dual cusps (Woodcock and Poston, 1976).

REFERENCES

Dubois, J-G. and Dufour, J-P. (1974) La machine à catastrophes. *Annales de l'Institut Henri Poincaré, XX,* 135-151.

Goodwin, A.N. (1971) Three dimensional pictures for Thom's parabolic umbilic. *Institute des Hautes Etudes Scientifiques Publications Mathematiques, 40,* 117-138.

Poston, T. (1975) Catastrophe machines. *Proceedings of the Symposium of the Battelle Memorial Institute.* Seattle: Seattle Research Center.

Poston, T. and Woodcock, A.E.R. (1973) Zeeman's catastrophe machine. *Proceedings of the Cambridge Philosophical Society, 74,* 217-226.

Stewart, I. and Woodcock, A.E.R. The reduced double swallowtail. To appear.

Thom, R. (1969) Topological Models in Biology. *Topology, 8,* 313-35.

Thom, R. (1972) *Stabilité Structurelle et Morphogenesis.* Reading: Benjamin Company, Inc.

Thom, R. (1975) *Structural Stability and Morphogenesis,* (D.W. Fowler, trans.). Reading: Benjamin Company, Inc.

Wasserman, G. (1975) *Proceedings of the Symposium of the Battelle Memorial Institute.* Seattle: Battelle Memorial Institute, Seattle Research Center.

Woodcock, A.E.R. (1974a) The simple cusp. In *The Geometrical Study of the Elementary Catastrophes,* Woodcock, A.E.R. and Poston, T. (New York: Springer-Verlag New York, Inc.)

Woodcock, A.E.R. (1974b) The swallowtail. In *The Geometrical Study of the Elementary Catastrophes,* Woodcock, A.E.R. and Poston, T. (New York: Springer-Verlag New York, Inc.)

Woodcock, A.E.R. (1974c) The butterfly. In *The Geometrical Study of the Elementary Catastrophes,* Woodcock, A.E.R. and Poston, T. (New York: Springer-Verlag New York, Inc.)

Woodcock, A.E.R. (1974d) The wigwam. In *The Geometrical Study of the Elementary Catastrophes,* Woodcock, A.E.R. and Poston, T. (New York: Springer-Verlag New York, Inc.)

Woodcock, A.E.R. (1974e) The star. In *The Geometrical Study of the Elementary Catastrophes,* Woodcock, A.E.R. and Poston, T. (New York: Springer-Verlag New York, Inc.)

Woodcock, A.E.R. (1974f). Cellular differentiation and catastrophe theory. *Annals of the New York Academy of Sciences, 231,* 60-76.

Woodcock, A.E.R. The geometry of the (1-3) and (3-1) stable catastrophes. To appear.

Woodcock, A.E.R. and Appel, A. Hidden-line elimination pictures of the umbilic catastrophes. To appear.

Woodcock, A.E.R. and Poston, T. (1976) A higher catastrophe machine. *Mathematical Proceedings of the Cambridge Philosophical Society, 79,* 343-350.

Woodcock, A.E.R. and Poston, T. (1974a) The cuspoids. In *The Geometrical Study of the Elementary Catastrophes,* Woodcock, A.E.R. and Poston, T. (New York: Springer-Verlag New York, Inc.)

Woodcock, A.E.R. and Poston, T. (1974b) The hyperbolic and elliptic umbilics. In *The Geometrical Study of the Elementary Catastrophes,* Woodcock, A.E.R. and Poston, T. (New York: Springer-Verlag New York, Inc.)

Woodcock, A.E.R. and Poston, T. (1974c) The parabolic umbilic. In *The Geometrical Study of the Elementary Catastrophes,* Woodcock, A.E.R. and Poston, T. (New York: Springer-Verlag New York, Inc.)

Woodcock, A.E.R. and Poston, T. (1974d) The reduced double cusp. In *The Geometrical Study of the Elementary Catastrophes,* Woodcock, A.E.R. and Poston, T. (New York: Springer-Verlag New York, Inc.)

Woodcock, A.E.R. and Poston, T. The geometry of the symbolic and anonymous umbilics. To appear.

Woodcock, A.E.R., Stewart, I. and Poston, T. The full unfolding of the double cusp. To appear.

Zeeman, E.C. (1972) A catastrophe machine. In *Towards a Theoretical Biology: Volume 4, Essays,* C.H. Waddington, ed. (Chicago: Aldine-Atherton, Inc.)

Zeeman, E.C. (1974) Primary and secondary waves in developmental biology. In *Lectures on Mathematics in the Life Sciences, Volume 7.* (Providence: American Mathematical Society).

Zeeman, E.C. (1976) Catastrophe theory. *Scientific American, 234,* 65-83.

CELLULAR COORDINATION IN EMBRYOGENESIS: SOME THEORETICAL CONSIDERATIONS

Introduction

Experimental investigations and theoretical analyses of embryogenesis have attempted to discover and describe the processes whereby a newly fertilized egg cell is transformed into an adult organism. These investigations have centered on the possible organismic, multicellular, cellular, cytoplasmic and genetic organization responsible for this transformation. Theoretical models, reflecting to some extent contemporary activity in the physical sciences and mathematics, have attempted to provide more or less precise mechanisms for these processes. The present paper reviews some of these studies and their contribution to an understanding of the nature of embryogenesis and the emergence of adult organisms.

Initial surgical experiments and the construction of rather naive physical models for developmental processes (reviewed in Russell, 1917; Huxley and deBeer, 1934; Thompson, 1942; Harrison, 1969) paved the way for the possibility of major advances in the study of embryological morphogenesis. Early expressions of the overriding importance of genetic participation in cytoplasmic epigenesis (see Wilson, 1928) were followed by numerous experiments, some using nuclear transplantation for instance, (King and Briggs, 1956; Gurdon and Woodland, 1969; Gurdon, 1974 for example) in which this role has been substantiated. These experiments and theoretical models (Britten and Davidson, 1969; Davidson and Britten, 1973; Crick, 1971) also suggest the crucial importance of close nuclear and cytoplasmic interactions in these processes.

Proposals that differential cellular adhesion (Townes and Holtfreter, 1955; Steinberg, 1963, 1970) and changes in the nature of the cellular boundary (Kosher and Searls, 1973; Hay, 1973; Hay and Meier, 1974) are responsible for the emergence of specialized cellular regions have been explored in a number of preparations. These studies also suggest the participation of "cellular recognition mechanisms" (Weiss, 1947; Roth, 1973) associated with the cell surface as important components of the developmental process.

Recent investigations (Iliano *et al.*, 1973; Lewin, 1973) implicate the participation of both cyclic nucleotides and insulin in a 'pleiotropic program' of growth and differentiation (Kram and Tomkins, 1973; Kram *et al.*, 1973).

The possibility that intercellular electrical coupling plays a role in the processes of cellular differentiation has been investigated in a number of preparations (Ito and Hori, 1966; Potter *et al.*, 1966; Bennett and Trinkaus, 1970; Palmer and Slack, 1970). While local membrane changes appear to occur during development, their nature and full importance does not seem to have been fully explored or established.

The proposal that gradients of some substances may participate actively in the regulation of embryogenetic processes has been the subject of extensive study (Morgan, 1905; Huxley and deBeer, 1934; Spemann, 1938; Child, 1941; Waddington, 1956, 1966; Goodwin, 1971). The participation of so-called 'organizer' regions in the processes of gastrulation (Spemann and Mangold, 1924; Nieuwkoop, 1973) may be seen as an extension of this suggestion. While much attention has been paid to detecting the substances participating in the establishment of the proposed gradients, widespread evidence for their existence, at least in animal tissue, seems to be lacking. Other theoretical models (Wall, 1973) utilize messenger ribonuclear protein in the establishment of physiological gradients supporting cellular coordination and subsequent morphogenesis. Extensive evidence supporting the participation of long-range electrical potential gradients in such processes has also not been clearly established (Slack and Wolpert, 1972). However, local electrical events may be important in determining intercellular activity (Slack and Warner, 1973).

In an attempt to circumvent the drawbacks of the classical gradient theories, the concept of positional information was promoted as a description for the processes providing long-range tissue coordination in developing embryos (Wolpert, 1969, 1971). Several models based on this principle seem to be reasonably successful in describing the emergence of differentiated regions in some preparations (Cooke, 1972, 1975a,b; Summerbell *et al.*, 1973; Webster, 1971; Lawrence *et al.*, 1972; Tickle *et al.*, 1975). However, it has been suggested (Nieuwkoop, 1973) that the use of such abstract labels as positional information may do little to advance the understanding of the nature of the basic mechanisms involved in embryogenesis. A model based on the precise discrimination of the relative phases of waves of some type within developing cellular tissue (Goodwin and Cohen, 1969; Goodwin, 1971; Cohen, 1971) has also been proposed as a basis for the coordination of cellular activities in embryogenesis.

Models based on reaction-diffusion mechanisms predict the production of standing concentration waves of so-called "morphogens" in cellular systems (Lotka, 1910; Turing, 1952; Othmer and Scriven, 1971, 1974). These schemes have been suggested as crucial components in the processes leading to the establishment of patterns in the emerging organism. Complex patterns have been realized in artificial, physico-chemical reactions (Zhabotinsky and Zaikin, 1973; Winfree, 1974).

The possibility that information theory, cybernetics and automata theory approaches might be useful in the analysis of developing embryonic tissues has received attention (Apter and Wolpert, 1965; Apter, 1966; Baer and Martinez, 1974). Recently models based on 'Developmental Algorithms' have also attempted to provide a theoretical framework for the description of embryonic processes (Lindenmayer, 1975). In this model, development is discussed in terms of growing arrays of finite automata with state transition functions providing instructions for changes within the system. These algorithmic processes strongly imply genetic participation and regulation in providing the information required by a developing cellular system.

In light of their possible hierarchical organization Pattee (1969, 1970, 1973) proposed new approaches to the investigation of biological systems. He suggested that this investigation may require the development of a new or at least more appropriate mathematical language for the description of biological processes before significant progress toward their understanding can be achieved.

Waddington (1975) pointed out that since Information- and Complexity-Theoretic approaches to an understanding of embryogenesis have met with mixed success, other approaches should be attempted. He suggested that an adequate description of embryogenesis can be achieved by considering epigenetic processes which explain the development of an adult organism in terms of the establishment of attracting trajectories (or 'chreods') on an "Epigenetic Landscape" (Waddington, 1940, 1956). The formulation of the epigenetic landscape itself appears to be a rather *ad hoc* representation both of the general nature of embryogenesis and of the mechanisms producing cellular growth and differentiation. Thom (1969, 1972, 1975) has suggested that a model based on catastrophe theory might provide a more adequate description of cellular behavior leading to morphogenesis. Catastrophe Theory (Thom, 1969; Woodcock, 1974) describes the nature of a generalized system in terms of the behavior of the stationary values of a potential energy function associated with that system. A surface representing these stationary values may be created by the loci of the maximum and minimum values of the energy function for particular values of the constraints on the system. Trajectories of a state point on that surface represent transitions generated in the system by changes in the nature or magnitude of the constraints, or control parameters, of the system. In this respect the 'catastrophe landscape' appears to be of greater importance than the 'epigenetic landscape', since it has emerged from a more rigorous mathematical treatment. In two independent studies, Woodcock (1974) and Zeeman (1974) have attempted to describe embryogenic processes in terms of models based on catastrophe theory. A direct consequence of the theory indicates that, in a system with up to four controlling parameters, only seven possible types of transitions between states are permitted (Thom, 1969). Zeeman (1974) suggested that cellular differentiation is triggered by the passage of a 'primary wave' of energy through the cellular mass. The primary wave is then followed by a 'secondary wave' of cellular activity which may instigate morphogenetic processes. This model has been extended to include the participation of a wave-front of rapid cellular change, a clock and the notion of positional information in the generation of patterns in embryonic tissue (Cooke and Zeeman, 1975). This theory appears, therefore, to be a hybrid of the earlier gradient theories with energy replacing the proposed morphogen required by those theories, and of ideas of positional information (Wolpert, 1969, 1971), the relative phase model of Goodwin and Cohen (1969) and of theories relying on regions of differential metabolism to support embryonic cellular differentiation (Rose, 1952).

It is clear that the embryogenic and morphogenic processes leading to the development of adult organisms are caused by a multitude of cellular biochemical reactions. Attempts at a completely comprehensive description of these processes appear to be doomed to failure at the outset due to the lack of precise information about local chemical concentrations, the nature of their interactions and participation in macromolecular, membrane and cellular organization, for example. However, recent investigations which implicate cyclic AMP, cyclic GMP and perhaps insulin (Robison *et al.*, 1968; Iliano *et al.*, 1973) as crucial factors in the regulation of some or all of the many biochemical processes leading to cellular growth or differentiation (Kram and Tomkins, 1973) suggest a way to offset these problems. These studies imply the possibility that the regulation or determination of the local concentrations of these chemicals may result in the system adopting particular patterns of biochemical activity. These different patterns of activity may, in turn, support the existence of overt differences in cellular morphology. This key role of the cyclic nucleotides in influencing many biochemical reactions may be compared to the action of the control parameters in determining the nature of the stationary states of the associated energy function in elementary catastrophe theory (Woodcock, 1974). It may also be possible to identify genetic processes whose influence, in regulating the nature of the cellular activity and its subsequent morphological expression could also resemble the activity of the control parameters in the elementary theory in determining the state of the cellular system. If such an identification can be made then the catastrophe manifold, and possible trajectories on the manifold under the influence of changes in the control parameters, may provide a qualitative description of the result of the cellular biochemical processes. This description may be an initial and probably imperfect formulation of the new language that seems to be required for the adequate description of very complex biological processes. Furthermore, the hierarchical organization of biological processes and their discontinuities at the molecular, cellular or organismic levels may suggest possible extensions of elementary catastrophe theory which could provide a more adequate description of these biological processes. In this context the theory, in attempting to describe processes at a particular level of the hierarchy, could employ control parameters and information appropriate to the level under investigation. For example, molecular organization could be described in terms of molecular energy and chemical bond angles (Gupta, Govil and Mishra, 1975), biochemical processes by thermodynamic free energy and substance concentrations (Henderson and Wang, 1972) and membrane processes by electrical potential energy and ionic conductivity (Hodgkin and Huxley, 1952). In this manner, local processes may be seen to be embedded within global processes or higher hierarchical levels, with each process described in terms of changes in the parameters of the system pertinent to that level. Woodcock (1974), however, has pointed out that difficulties may arise both in the description of an adequate energy function required by elementary catastrophe theory and in the necessity of local continuity or differentiability of this energy function in some circumstances.

The exhibition of oscillations in biological systems strongly suggests that limit cycle behavior must be a part of any possible model of such systems. The simple cusp does support potentially oscillatory behavior (Zeeman, 1972) since some trajectories resemble those of the Van der Pol oscillator (Van der Pol, 1926). The recently announced (1-3) and (3-1) stable catastrophe theory (Wasserman, 1975 to appear; Woodcock to appear) may provide a more adequate description of some developmental processes.

The bifurcation of limit cycles can occur in ways described by the elementary catastrophes; the relation of the Duffing equation to the cusp catastrophe (Holmes and Rand, 1976) provides an important example. However, other phenomena such as the Hopf bifurcation can occur in stable systems, and the appropriate topological methods, those of general Dynamical Systems Theory, are less completely developed.

Thom (1975) observed that complicated changes in the nature of a system (for example, the emergence of a limit cycle by Hopf bifurcation) cannot easily be described in terms of elementary catastrophe theory, and suggested that an extension of the theory in this direction may be a difficult matter. He has proposed that so-called generalized catastrophes may occur when a limit cycle is annihilated. Under this condition each local region of the dynamical system would appear to behave as if it was undergoing a local elementary catastrophe transition. Therefore, a description of embryogenesis in terms of elementary catastrophe theory alone may be misleading; more comprehensive methods may be, almost certainly will be, necessary before a complete and adequate description of embryogenesis can be attempted on the basis of catastrophe theory. Even this is a 'semilocal' approach which ultimately should give way to a fusion of topological ideas, such as structural stability, with the holistic techniques of Global Analysis.

Finally, the theoretical analysis of embryogenesis should reflect the importance of both genetic and cytoplasmic processes in the regulation of synthetic biochemical activity. It is possible, for example, that the existence of overt gradients and other localized property differences within developing tissue may be a consequence of cellular growth and differentiation and not their cause. Models based on these types of differences in property, therefore, would be misleading if considered without the crucial involvement of genetic processes in cellular development and differentiation.

REFERENCES

Apter, M.J. (1966) *Cybernetics and Development*. Oxford: Pergamon Press, Ltd.

Apter, M.J. and Wolpert, L. (1965) Cybernetics and development I. Information theory. *Journal of Theoretical Biology, 8,* 244-257.

Baer, R.M. and Martinez, H.M. (1974) Automata and biology. *Annual Review of Biophysics and Bioengineering, 3,* 255-291.

Bennett, M.V.L. and Trinkaus, J.P. (1970) Electrical coupling between embryonic cells by way of extracellular space and specialized junctions. *Journal of Cell Biology, 44,* 592-610.

Britten, R.J. and Davidson, E.N. (1969) Gene regulation for higher cells: A theory. *Science, 165,* 349-357.

Child, C.M. (1941) *Patterns and Problems of Development*. Chicago: University of Chicago Press.

Cohen, M.H. (1971) Models for the control of development. *Symposia of the Society for Experimental Biology, 25,* 455-476.

Cooke, J. (1972) Properties of the primary organization field in the embryo of *Xenopus laevis*. II. Positional information for axial organization in embryos with two head organizers. *Journal of Embryology and Experimental Morphology, 28,* 27-46.

Cooke, J. (1975a) Control of somite number during morphogenesis of a vertebrate, *Xenopus laevis*. *Nature, 254,* 196-199.

Cooke, J. (1975b) The emergence and regulation of spatial organization in early animal development. *Annual Review of Biophysics and Bioengineering, 4,* 185-217.

Cooke, J. and Zeeman, E.C. (1975) A clock and wavefront model for control of the number of repeated structures during animal morphogenesis. Preprint. Coventry: University of Warwick.

Crick, F.H.C. (1971) General model for the chromosomes of higher organisms. *Nature, 234,* 25-27.

Davidson, E.H. and Britten, R.J. (1973) Organization, transcription and regulation in the animal genome. *Quarterly Review of Biology, 48,* 565-613.

Goodwin, B.C. (1971) A model of early amphibian development. *Symposia of the Society for Experimental Biology, 25,* 417-428.

Goodwin, B.C. and Cohen, M.H. (1969) A phase shift model for the spatial and temporal organization of developing systems. *Journal of Theoretical Biology, 25,* 49-107.

Gupta, S.P., Govil, G. and Mishra, R.K. (1975) Molecular orbital studies on the conformation of phospholipids II. Preferred conformations of hydrocarbon chains and molecular organization in biomembranes. *Journal of Theoretical Biology, 51,* 13-34.

Gurdon, J.B. (1974) *The Control of Gene Expression in Animal Development.* Cambridge: Harvard University Press.

Gurdon, J.B. and Woodland, H.R. (1969) The influence of the cytoplasm on the nucleus during cell differentiation with special reference to RNA synthesis during amphibian cleavage. *Proceedings of the Royal Society, B173,* 99-111.

Harrison, R.G. (1969) Cellular differentiation and internal environment. In *Organization and Development of the Embryo,* S. Wilens, ed. (New Haven: Yale University Press).

Hay, E.D. (1973) The origin and the role of collagen in the embryo. *American Zoologist, 13,* 1085-1107.

Hay, E.D. and Meier, S. (1974) Glycosaminoglycan synthesis by embryonic inductors: Neural tube, notochord and lens. *Journal of Cell Biology, 62,* 889-898.

Henderson, R. and Wang, J.H. (1972) Catalytic configurations. *Annual Review of Biophysics and Bioengineering, 1,* 1-26.

Hodgkin, A.L., and Huxley, A.F. (1952) Currents carried by sodium and potassium ions through the membrance of the giant axon of *Loligo. Journal of Physiology, 116,* 449-472.

Holmes, P.J. and Rand, D.A. (1976) The bifurcation of Duffing's equations: An application of an approach to catastrophe theory. *Journal of Sound and Vibration, 44,* 237-253.

Huxley, T.E. and deBeer, G. (1934) *The Elements of Experimental Embryology.* New York: The Macmillan Company.

Iliano, G., Tell, G.P.E., Siegel, M.I. and Cuatrecasas, P. (1973) Guanosine 3':5'-cyclic monophosphate and the action of insulin and acetylcholine. *Proceedings of the National Academy of Sciences of the United States of America, 70,* 2443-2447.

Ito, S. and Hori, N. (1966) Electrical characteristics of *Triturus* egg cells during cleavage. *Jorunal of General Physiology, 49,* 1019-1027.

King, T.J. and Briggs, R. (1956) Serial transplantation of embryonic nuclei. *Cold Spring Harbor Symposia on Quantitative Biology, 21,* 271-290.

Kosher, R.A. and Searls, R.L. (1973) Sulfated mucopolysaccharide synthesis during the development of *Rana pipiens. Developmental Biology, 32,* 50-68.

Kram, R., Mamont, P. and Tomkins, G.M. (1973) Pleiotypic control by adenosine 3':5' cyclic monophosphate: A model for growth control in animal cells. *Proceedings of the National Academy of Sciences of the United States of America, 70*, 1432-1436.

Kram, R. and Tomkins, G.M. (1973) Pleiotypic control by cyclic AMP: Interaction with cyclic GMP and possible role of microtubules. *Proceedings of the National Academy of Sciences of the United States of America, 70*, 1659-1663.

Lawrence, P.A., Crick, F.H.C. and Munro, M. (1972) A gradient of positional information in an insect *Rhodnius*. *Journal of Cell Science, 11*, 815-53.

Lewin, R. (1973) New role for insulin. *New Scientist*, 690-692.

Lindenmayer, A. (1975) Developmental algorithms for multicellular organisms: A survey of L-systems. *Journal of Theoretical Biology, 54*, 3-22.

Lotka, A.J. (1910) Contribution to the theory of periodic reactions. *Journal of Physical Chemistry, 14*, 271-274.

Morgan, T.H. (1905) 'Polarity' considered as a phenomenon of gradation of material. *Journal of Experimental Zoology, 2*, 495.

Nieuwkoop, P.D. (1973) The 'Organization Center' of the amphibian embryo: its origin, spatial organization and morphogenetic action. *Advances in Morphogenesis, 10*, 1-39.

Othmer, H.G. and Scriven, L.E. (1971) Instability and dynamic pattern in cellular networks. *Journal of Theoretical Biology, 32*, 507-537.

Othmer, H.G. and Scriven, L.E. (1974) Non linear aspects of dynamic patterns in cellular networks. *Journal of Theoretical Biology, 43*, 83-112.

Pattee, H.H. (1969) How does a molecule become a message? *Symposia of the Society for Developmental Biology, 28*, 1-16.

Pattee, H.H. (1970) The problem of biological hierarchy. In *Towards a Theoretical Biology: Volume 3, Drafts*, C.H. Waddington, ed. (Chicago: Aldine-Atherton, Inc.)

Pattee, H.H. (1973) The physical basis and origin of hierarchical control. In *Hierarchy Theory, The Challenge of Complex Systems*, H.H. Pattee, ed. (New York: G. Braziller, Inc.)

Palmer, J.F. and Slack, C. (1970) Some bioelectric parameters of early *Xenopus* embroyos. *Journal of Embryology and Experimental Morphology, 24*, 535-554.

Potter, D.D., Furshpan, E.J. and Lennox, E.S. (1966) Connections between cells of the developing squid as revealed by electrophysiological methods. *Proceedings of the National Academy of Sciences of the United States of America, 55*, 328-336.

Robison, G.A., Butcher, R.W., and Sutherland, E.W. (1968) Cyclic AMP. *Annual Review of Biochemistry, 37,* 149-174.

Rose, S.M. (1952) A hierarchy of self-limiting reactions as the basis of cellular differentiation and growth control. *American Naturalist, 86,* 337-354.

Roth, S.A. (1973) A molecular model for cell interactions. *Quarterly Review of Biology, 48,* 541-563.

Russell, E.S. (1917) *Form and Function: A Contribution to the History of Animal Morphology.* New York: E.P. Dutton Co.

Slack, C. and Warner, A.E. (1973) Intracellular and intercellular potentials in the early amphibian embryo. *Journal of Physiology, 232,* 313-330.

Slack, C. and Wolpert, L. (1972) Absence of intracellular gradients in amphibian embryos. *Nature, New Biology, 236,* 153-155.

Spemann, H. (1938) *Embryonic Development and Induction.* New Haven: Yale University Press.

Spemann, H. and Mangold, H. (1924) Über Weckung organisatovischer Fähigkeiten durch Verpflanzung in organisatovische umgebung. *Wilhelm Roux Archiv für Entwicklungsmechanik der Organismen, 100,* 599-638.

Steinberg, M.S. (1963) Reconstruction of tissues by dissociated cells. *Science, 141,* 401-408.

Steinberg, M.S. (1970) Does differential adhesion govern self-assembly processes in histogenesis? Equilibrium configurations and the emergence of a hierarchy among populations of embryonic cells. *Journal of Experimental Zoology, 173,* 395-434.

Summerbell, D., Lewis, J.H., and Wolpert, L. (1973) Positional information in chick limb morphogenesis. *Nature, 244,* 492-496.

Thom, R. (1969) Topological models in biology. *Topology, 8,* 313-335.

Thom, R. (1972) *Stabilité Structurelle et Morphogénèse.* Reading: Benjamin Company, Inc.

Thom, R. (1975) *Structural Stability and Morphogenesis.* (D.W. Fowler, trans.) Reading: Benjamin Company, Inc.

Thompson, D'Arcy W. (1942) *On Growth and Form.* Cambridge: Cambridge University Press.

Tickle, C., Summerbell, D. and Wolpert, L. (1975) Positional signalling and specification of digits in chick limb morphogenesis. *Nature, 254,* 199-202.

Townes, P.L. and Holtfreter, J. (1955) Directed movements and selective adhesion of embryonic amphibian cells. *Journal of Experimental Zoology, 128*, 53-120.

Turing, A.M. (1952) The chemical basis of morphogenesis. *Philosophical Transactions of the Royal Society, B237*, 37-72.

Van der Pol, B. (1926) On relaxation oscillators. *Philosophical Magazine, 2*, 978-992.

Waddington, C.H. (1940) *Organizers and Genes*. Cambridge: Cambridge University Press.

Waddington, C.H. (1956) *Principles of Development*. London: Allen and Unwin, Ltd.

Waddington, C.H. (1966) Fields and gradients. *Symposia of the Society for Developmental Biology, 25*, 105-124.

Waddington, C.H. (1975) *The Evolution of an Evolutionist*. Ithaca: Cornell University Press.

Wall, R. (1973) Physiological gradients in development - a possible role for messenger ribonucleoprotein. *Advances in Morphogenesis, 10*, 41-114.

Wasserman, G. (1975) *Proceeding of the Symposium of the Battelle Memorial Institute*. Seattle: Battelle Memorial Institute, Seattle Research Center.

Webster, G. (1971) Morphogenesis and pattern formation in hydroids. *Biological Reviews, 46*, 1-46.

Weiss, P. (1947) The problem of specificity in growth and development. *Yale Journal of Biology and Medecine, 19*, 235-278.

Wilson, E.B. (1928) *The Cell in Development and Heredity*. (Third Edition). New York: The Macmillan Company.

Winfree, A.T. (1974) Rotating chemical reactions. *Scientific American, 230*, 82-95.

Wolpert, L. (1969) Positional information and the spatial pattern of cellular differentiation. *Journal of Theoretical Biology, 25*, 1-47.

Wolpert, L. (1971) Positional information and pattern formation. *Current Topics in Developmental Biology, 6*, 183-224.

Woodcock, A.E.R. (1974) Cellular differentiation and catastrophe theory. *Annals of the New York Academy of Sciences, 231*, 60-76.

Woodcock, A.E.R. (1-3) and (3-1) stable catastrophes: a description of cellular differentiation? To appear.

Woodcock, A.E.R. The geometry of the (1-3) and (3-1) stable catastrophes. To appear.

Zeeman, E.C. (1972) Differential equation for the heartbeat and nerve impulse. In *Towards a Theoretical Biology: Volume 4, Essays*, C.H. Waddington, ed. (Chicago: Aldine-Atherton, Inc.)

Zeeman, E.C. (1974) Primary and secondary waves in developmental biology. In *Lectures on Mathematics in the Life Sciences, Volume 7*. (Providence: American Mathematical Society).

Zhabotinsky, A.M. and Zaikin, N. (1973) Autowave processes in a distributed chemical system. *Journal of Theoretical Biology, 40*, 45-61.

MODELS OF THE NERVE IMPULSE

Studies of a passive iron wire model (Lillie, 1925, 1936; Bonhoeffer, 1948) have attempted to provide an insight into the nature of the nerve impulse. These studies seemed to indicate that local circuit currents were involved in the longitudinal propagation of activity. The characteristic features of threshold, transmission of activity, refractory states, and the tendency toward rhythmic reactions exhibited in both living and artificial systems suggested that these properties are in some way interrelated (Bonhoeffer, 1948). To describe this behavior Bonhoeffer (1948) proposed equations of the form

$$\dot{x} = f(x,y), \quad \dot{y} = g(x,y), \tag{1}$$

where x is the degree of activation and y the refractoriness. He showed that stimulating currents just above Rheobase produced single pulses of activity while higher current densities generated rhythmic activity.

In an analysis of simple oscillatory systems, Van der Pol (1926) discussed the nature of triode vacuum tube oscillations in terms of solutions of the equation

$$\ddot{v} + \kappa\dot{v} + \omega^2 v = 0, \tag{2}$$

where v is the electrical potential and κ and ω are constants. Van der Pol (1926) modified this equation to incorporate a damping term dependent on v^2, i.e.

$$\ddot{v} + \varepsilon(1 - v^2)\dot{v} + v = 0 \tag{3}$$

and then plotted phase portraits of $\dot{v}$ against v for various values of the damping coefficient ε. These portraits exhibit periodic behavior with points spiralling toward a limit cycle. Van der Pol and Van der Mark (1928) extended the notion of oscillations and incorporated them in a model of cardiac behavior.

The Nerve Membrane at Rest

The electrophysiology of the nerve membrane , in particular that of the squid giant axon, has been the subject of extensive study (see Hodgkin and Huxley, 1952a,b). The large diameter of this nerve cell facilitates relatively easy access to the cytoplasm by means of micro-electrodes and other techniques. Chemical analysis (Hodgkin and Huxley, 1952a; Hodgkin, 1964) has revealed that the intracellular environment of the resting nerve has a relatively high potassium and low sodium and chloride ionic concentration in relation to the extracellular fluid. The potential difference across the nerve membrance at rest may be computed from the Goldman equation

$$\text{Resting Potential} = \frac{RT}{ZF}\ell n\left(\frac{P_K[K]_i + P_{Na}[Na]_i + P_{Cl}[Cl]_o}{P_K[K]_o + P_{Na}[Na]_o + P_{Cl}[Cl]_i}\right) \tag{4}$$

where the P_i's are ionic permeabilities and $[\ ]_i$ and $[\ ]_o$ represent the ionic concentrations inside and outside the nerve membrane (Goldman, 1943; Hodgkin & Katz, 1949). Calculations using the Nernst equation

$$\text{Ionic Equilibrium Potential} = 58 \log_{10} \frac{[\text{ion}]\text{outside}}{[\text{ion}]\text{inside}} \tag{5}$$

have suggested that the potassium ionic distribution is approximately in equilibrium, supported by the observed membrane potential. The nerve membrane at rest has a finite potassium conductivity (Hodgkin and Huxley, 1952a).

The Nerve Action Potential

Intracellular electrical recordings (Hodgkin and Huxley, 1952a) have demonstrated that the action potential consists of a transient reversal of electrical polarity from -60 mv (inside negative) to about 40 to 50 mv (inside positive).

The nature of ionic relations before and during the nerve action potential has been investigated by the technique of voltage clamping (Cole, 1949; Hodgkin and Huxley, 1952a; Hodgkin, Huxley and Katz, 1952). In these experiments the membrane potential was changed from rest, held fixed, and then the membrane current was measured with different extracellular ionic environments. (For example, sodium chloride was replaced by choling chloride in some experiments .)

The ionic conductivities of the membrane are given by

$$g_{Na} = \frac{I_{Na}}{V-V_{Na}}, \quad g_{k} = \frac{I_{K}}{V-V_{K}}, \quad g_{Cl} = \frac{I_{Cl}}{V-V_{Cl}}, \tag{6}$$

where g_i is the conductivity, I_i the ionic current, V_i the equilibrium potential and V the (clamped) membrane potential. By using these relations in conjunction with the voltage clamp technique, Hodgkin and Huxley (1952a) were able to plot membrane ionic current as a function of membrane potential. They demonstrated, in sodium replacement experiments, that the early component of the action potential was due to a fast inrush of sodium which was followed by a potassium outflux. This corresponded to an initial, transient increase of sodium membrane conductance and a subsequent increase in potassium membrane conductance. The sodium conductance rapidly returned to zero while the potassium conductance returned more slowly to a small, but measureable value. Hodgkin and Huxley (1952b) also investigated the electrical properties of the nerve membrane close to the threshold for the generation of the nerve action potential.

From their voltage clamp data, Hodgkin and Huxley (1952b) devised a mathematical model to describe the electrical behavior of the nerve membrane. They suggested that the cooperative activity of "four bound particles" is required for potassium ionic flow. If n is the probability that one particle is in the right place to permit the flow of potassium ions, the potassium ionic conductance is given by

$$g_K = \bar{g}_K n^4 \qquad (7)$$

where $\bar{g}_K$ is the maximum potassium conductance and

$$\frac{dn}{dt} = \alpha_n (1-n) - \beta_n n \qquad (8)$$

describes the time dependence of the n parameter. Furthermore, the coherent activity of three particles, each with a positional probability m, is required for the flow of sodium ions. An event of probability $(1-h)$ blocks this transport. Hence the sodium conductivity is given by

$$g_{Na} = \bar{g}_{Na} m^3 h \qquad (9)$$

where $\bar{g}_{Na}$ is the maximum sodium conductance. The equations

$$\frac{dm}{dt} = \alpha_m (1-m) - \beta_m m, \quad \frac{dh}{dt} = \alpha_h (1-h) - \beta_h h \qquad (10)$$

describe the time-dependence of the m and h parameters; the terms α and β depend on temperature, calcium concentration and membrane potential (Hodgkin and Huxley, 1952b).

The complete membrane ionic current density (Hodgkin and Huxley,

1952b; Hodgkin, 1964) is given by

$$I_m = (V - V_K)\bar{g}_K n^4 + (V - V_{Na})\bar{g}_{Na} m^3 h + (V - V_L)\bar{g}_L + c\,\frac{\partial V}{\partial t} \qquad (11)$$

where L represents the participation of other ions in the membrane current, c is the membrane capacity and $\frac{\partial V}{\partial t}$ the rate of change of membrane potential.

By using the equation for membrane current (Hodgkin and Huxley, 1952b; Hodgkin, 1964)

$$I_m = \frac{a}{2R}\frac{\partial^2 V}{\partial x^2}, \qquad (12)$$

the form and velocity of the propagated action potential can be determined. In equation (12), a is the radius, R is the resistivity of the axon and x is the distance along the axon. Combining (11) and (12) gives

$$\frac{a}{2R}\frac{\partial^2 V}{\partial x^2} = c\,\frac{\partial V}{\partial t} + (V - V_K)\bar{g}_K n^4 + (V - V_{Na})\bar{g}_{Na} m^3 h + (V - V_L)\bar{g}_L\,. \qquad (13)$$

For a continuous nerve fibre propagating an impulse at constant velocity θ, $x = -\theta t$ and equation (12) becomes

$$\frac{a}{2R\theta^2}\frac{d^2 V}{dt^2} = c\,\frac{dV}{dt} + (V - V_K)\bar{g}_K n^4 + (V - V_{Na})\bar{g}_{Na} m^3 h + (V - V_L)\bar{g}_L\,. \qquad (14)$$

Hodgkin and Huxley (1952b) attempted to solve equation (13) numerically.

The complexity of equation (13) and the difficulty experienced by Hodgkin and Huxley in obtaining a solution stimulated the introduction of a simpler version of the equation by Nagumo *et al.*, (1962). The Nagumo equation is

$$\frac{\partial v}{\partial t} = \frac{\partial^2 v}{\partial x^2} - f(v) - w$$

$$\frac{\partial w}{\partial t} = bv, \quad b > 0 . \tag{15}$$

Rinzel (1973) has determined all the periodic and pulse travelling wave solutions of the Nagumo equation and has analyzed their stability. Of the two types of travelling pulse solutions which he described, the one which resembled the nerve impulse is apparently stable while the other, a slower wave solution, appears to be unstable.

FitzHugh (1960, 1961) attempted a topological analysis of the non-linear differential equations which Hodgkin and Huxley (1952b) proposed as a description of the nerve impulse. An initial, qualitative investigation of the equations considered the critical paths, singular points, separatrices and limit cycles determining the behavior or trajectory of a 'phase point' on a 'phase space'. In an analysis of a reduced version of the Hodgkin and Huxley equations, FitzHugh (1960) demonstrated the existence of equilibrium, steady state, and saddle points on the phase plane. He also investigated the influence of the intensity of both cathodal and anodal stimulation on the trajectory of the phase point, and discussed the nature of the saddle point, its occurrence in the 'reduced' equations and its implication for the threshold of excitation. He pointed out that analysis of the complete system of Hodgkin-Huxley equations suggests that there is "no saddle point" and that "any intermediate response between 'all' and 'none' is obtainable by an accurate enough adjustment of stimulus intensity" (FitzHugh, 1960).

In the following year FitzHugh (1961) proposed that an extension of the Van der Pol (1926) analysis of relaxation oscillations would aid in the analysis of the Hodgkin and Huxley (1952b) nerve impulse equations. He suggested that the original Van der Pol equation

$$\ddot{x} + c(x^2 - 1)\dot{x} + x = 0 \tag{16}$$

be modified by suitable transformation to

$$\dot{x} = c(y + x - \frac{x^3}{3}) \tag{17}$$

where $\dot{y} = -x/c$. Further modification of (17) produced the equation

$$\dot{x} = c(y + x - \frac{x^3}{3} + z) \tag{18}$$

where

$$\dot{y} = -(x - a - by)/c, \tag{19}$$

a,b are constants and z is the stimulus intensity. The constraints

$$1 - 2b/3 < a < 1, \quad 0 < b < 1, \quad b < c^2 \tag{20}$$

guarantee the existence of one stable singular point. FitzHugh (1961) examined this model, the Bonhoeffer-Van der Pol (BVP) model, in terms of the stability of the resting singular point of the system. Analog computations revealed the existence of single pulsatile activities or of oscillatory behavior for certain values of the coefficients a,b,c and z. FitzHugh pointed out that the BVP model incorporates the basic dynamic interrelationships among the variables responsible for the properties of threshold, refractoriness and impulse trains. FitzHugh noted that the precise algebraic form of the equations is not important in such an analysis, since other equations have similar properties.

In an extension of the Van der Pol-FitzHugh phase portrait type of analysis, Zeeman (1972) has discussed the behavior of the processes which generate the nerve impulse in terms of a model based on catastrophe theory. In this model 'fast' and 'slow' equations govern the movement of the system towards equilibrium conditions after an initial perturbation. In particular the equations

$$\begin{aligned} \varepsilon\dot{x} &= -(x^3 - x + b) \\ \dot{b} &= x \end{aligned} \tag{21}$$

describe the Van der Pol Oscillator (Zeeman, 1972). This system has two 'attractor states' and one 'repellor state'.

Catastrophe theory describes the behavior of a system in terms of changes in the nature of an associated potential energy function. The maximum and minimum values of the energy function represent stationary states of the system and are determined by the relative magnitudes of the control parameters (Woodcock and Poston, 1974). Changes in the magnitudes of these parameters will, in general, cause a change in the nature of these stationary states. In applying the theory to describe the behavior of a given system the control parameters must be identified with variables whose action regulates the behavior of the system. In the case of the nerve impulse, Zeeman (1972) chose the potassium ionic conductivity and the membrane potential as the control parameters (a and b), and the sodium ionic conductivity as the 'behavior parameter' (x) of the system. More precisely, he gave the following relationships:

x: Sodium Conductivity $\equiv g_{Na} = (4[x + 0.5]_-)^2$ mmho/cm^2

a: Potassium Conductivity $\equiv g_K = 2.38[a + 0.5]_+$ mmho/cm^2

b: Membrane Potential $\equiv V = (20b + 16$ mv)

where

$$[y]_+ = \begin{cases} y, & y \geq 0 \\ 0, & y < 0 \end{cases}, \qquad [y]_- = \begin{cases} 0, & y \geq 0 \\ y, & y < 0 \end{cases}.$$

Zeeman (1972) also used

Chlorine Conductivity $\equiv g_{Cl} = 0.15$ mmho/cm^2 (Hodgkin, 1951)

Membrane Capacity $\equiv 1\mu$ Farad/cm^2 (Hodgkin and Huxley, 1952b)

in his computations. The local nerve impulse equations were defined in terms of the time differentials of functions of the membrane potential and the sodium and potassium conductivities,

$$\dot{x} = -1.25(x^3 + ax + b) \tag{22}$$

$$\dot{a} = (x + 0.06(a + 0.5))(x - 1.5a - 1.67)(0.054(b - 0.8)^2 + 0.75) \tag{23}$$

$$\dot{b} = -g_K(b + 1.4) - g_{Na}(b - 4.95) - g_{Cl}(b - 0.15) \tag{24}$$

and trajectories of the state point on the (a,b) plane may be found in Zeeman (1972, Figure 25).

A close inspection of Zeeman's picture of the state point trajectories reveals the possible existence of two or perhaps three separatrices or regions in which solutions of equations (22)-(24) converge. In this case, this figure therefore resembles those described earlier by FitzHugh (1961, see Figure 1 for example). If the use of a catastrophe theory model is an appropriate way to describe nerve impulse activity, a catastrophe with a higher-dimensional control space than that of the simple cusp might be required to form the basis of a completely satisfactory model (Woodcock and Poston, 1974). Indeed, the role of calcium and chloride ions in the processes of the nerve impulse (Fatt and Ginsborg, 1958) under some conditions strongly implies that a model using only two control parameters and therefore based on the simple cusp catastrophe is too simplistic an approach to the problem. A computer analysis of the Zeeman equation also reveals, even using the simple cusp catastrophe, that the Hodgkin and Huxley (1952a,b) voltage clamp data is better approximated if some of the coefficients suggested by Zeeman (1974) are altered.

A discussion of these matters and a fuller analysis of the Zeeman equations will appear elsewhere (Woodcock, to appear).

The models of the nerve impulse reviewed in this paper provide different types of ways to illustrate the nature of the processes underlying nerve impulse activity. The use of differential equations (Hodgkin and Huxley, 1952b) provides a quantitative description, while non-linear mechanics (FitzHugh, 1961) or catastrophe theory (Zeeman, 1972) provides a qualitative description of the behavior of excitable neurophysiological systems. The use of these and related approaches has suggested, and probably will suggest, possible experiments to further explore this behavior.

REFERENCES

Appleton, E.V. and Van der Pol, B. (1922) On a type of oscillation-hysteresis in a simple triode generator. *Philosophical Magazine, 43,* 177-193.

Bonhoeffer, K.F. (1948) Activation of passive iron as a model for the excitation of nerve. *Journal of General Physiology, 32,* 69-91.

Cole, K.S. (1949) Dynamic electrical characteristics of the squid axon membrane. *Archives des Sciences Physiologiques, 3,* 253-258.

Fatt, P. and Ginsborg, B.L. (1958) The ionic requirements for the production of action potentials in crustacean muscle fibres. *Journal of Physiology,142,* 516-543.

FitzHugh, R. (1960) Thresholds and plateaus in the Hodgkin-Huxley nerve equations. *Journal of General Physiology, 43,* 867-896.

FitzHugh, R. (1961) Impulses and physiological states in the theoretical models of nerve membrane. *Biophysical Journal, 1,* 445-466.

Goldman, D.E. (1943) Potential impedance, and rectification in membranes. *Journal of General Physiology, 27,* 37-60.

Hodgkin, A.L. (1951) The ionic basis for electrical activity in nerve and muscle. *Biological Reviews, 26,* 339-409.

Hodgkin, A.L. (1964) *The Conduction of the Nerve Impulse.* Liverpool: Liverpool University Press.

Hodgkin, A.L. and Huxley, A.F. (1952a) Currents carried by sodium and potassium ions through the membrance of the giant axon of *Loligo.* *Journal of Physiology, 116,* 449-472.

Hodgkin, A.L. and Huxley, A.F. (1952b) A quantitative description of membrane current and its application to conduction and excitation in nerve. *Journal of Physiology, 117,* 500-544.

Hodgkin, A.L., Huxley, A.F. and Katz, B. (1952) Measurement of current-voltage relations in the membrane of the giant axon of *Loligo.* *Journal of Physiology, 116,* 424-448.

Hodgkin, A.L. and Katz, B. (1949) The effect of sodium ions on the electrical activity of the giant axon of the squid. *Journal of Physiology, 108,* 37-77.

Lillie, R.S. (1925) Factors affecting transmission and recovery in the passive iron nerve model. *Journal of General Physiology, 7,* 473-507.

Lillie, R.S. (1936) The passive iron wire model of protoplasmic and nervous transmission and its physiological analogues. *Biological Reviews, 11,* 181-209.

Nagumo, J., Arimoto, S. and Yoshizawa, S. (1962) An active pulse transmission line simulating nerve axon. *Proceedings of the Institute of Radio Engineers, 50,* 2061-2070.

Rinzel, J. (1973) Travelling wave solutions of a nerve conduction equation. Ph.D. Dissertation. New York: New York University.

Van der Pol, B. (1926) On 'Relaxation-Oscillations'. *Philosophical Magazine, 2,* 978-992.

Van der Pol, B. and Van der Mark, J. (1928) The heartbeat considered as a relaxation oscillation and an electrical model of the heart. *Philosophical Magazine, 6,* 763-775.

Woodcock, A.E.R. The nerve impulse and the simple cusp catastrophe - an analysis of a model. To appear.

Woodcock, A.E.R. and Poston, T. (1974) *The Geometrical Study of the Elementary Catastrophes.* New York: Springer-Verlag New York, Inc.

Zeeman, E.C. (1972) Differential equations for the heartbeat and nerve impulse. In *Towards a Theoretical Biology: Volume 4, Essays,* C.H. Waddington, ed. (Chicago: Aldine-Atherton, Inc.)

This series aims to report new developments in biomathematics research and teaching – quickly, informally and at a high level. The type of material considered for publication includes:

1. Preliminary drafts of original papers and monographs
2. Lectures on a new field, or presenting a new angle on a classical field
3. Seminar work-outs
4. Reports of meetings, provided they are
   a) of exceptional interest and
   b) devoted to a single topic.

Texts which are out of print but still in demand may also be considered if they fall within these categories.

The timeliness of a manuscript is more important than its form, which may be unfinished or tentative. Thus, in some instances, proofs may be merely outlined and results presented which have been or will later be published elsewhere. If possible, a subject index should be included. Publication of Lecture Notes is intended as a service to the international scientific community, in that a commercial publisher, Springer-Verlag, can offer a wider distribution to documents which would otherwise have a restricted readership. Once published and copyrighted, they can be documented in the scientific literature.

Manuscripts

Manuscripts should comprise not less than 100 and preferably not more than 500 pages.
They are reproduced by a photographic process and therefore must be typed with extreme care. Symbols not on the typewriter should be inserted by hand in indelible black ink. Corrections to the typescript should be made by pasting the amended text over the old one, or by obliterating errors with white correcting fluid. Authors receive 75 free copies and are free to use the material in other publications. The typescript is reduced slightly in size during reproduction; best results will not be obtained unless the text on any one page is kept within the overall limit of 18 x 26.5 cm (7 x 10½ inches). The publishers will be pleased to supply on request special stationery with the typing area outlined.

Manuscripts in English, German or French should be sent to Dr. Simon Levin, Center for Applied Mathematics, Olin Hall, Cornell University Ithaca, NY 14850/USA or directly to Springer-Verlag Heidelberg.

Springer-Verlag, Heidelberger Platz 3, D-1000 Berlin 33
Springer-Verlag, Neuenheimer Landstraße 28–30, D-6900 Heidelberg 1
Springer-Verlag, 175 Fifth Avenue, New York, NY 10010/USA

ISBN 3-540-08351-0
ISBN 0-387-08351-0